Military in the Rear View Mirror:

Mental Health and Wellness

in Post-Military Life

Duane K. L. France, MA, MBA, LPC

i

MILITARY IN THE REAR VIEW MIRROR

Published By:

NCO Historical Society

P.O. Box 1341

Temple, TX 76503

www.ncohistory.com

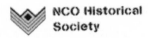 **NCO Historical Society**

The content of this book has been previously published in a digital format on the Head Space and Timing blog, located at www.veteranmentalhealth.com

Cover Design by Extended Imagery

The author of this book is a Mental Health Counselor licensed to practice in the state of Colorado. The thoughts, ideas, musings, and posts in this book come from his military experience, professional experience, and personal opinions. They do NOT, however, represent professional advice. While he is a Mental Health Counselor, he is not YOUR Mental Health Counselor, and the guidance in this work should not be considered a substitute for working with a licensed clinical mental health provider. The opinions expressed here are his own, and in no way should be seen as reflection of his agency, his profession, or any professional associations that he is connected with.

ISBN: 978-0-9963181-5-0

Published in the United States of America

1st Edition

To those who sacrifice:

Service Members,

Veterans,

Their Families,

And those who dedicate their life's work

to helping those who have the military in the rear view mirror

CONTENTS

Foreword

Duane France's newest book, *Military in the Rear-View Mirror*, spins on the axis of this statement: *"The only veteran whose life you can save tomorrow is yours."* This may seem like an odd statement for a combat veteran who has since become a licensed mental health provider…even more interesting perhaps given that Duane is my co-host for the nationally disseminated podcast "Seeking the Military Suicide Solution." Yet, as fellow healers, and thought leaders in the suicide prevention field, Duane and I agree on this principle.

We empower people by putting them in the driver's seat. I've found that veterans and first responders are often concerned that therapy will be disempowering. To some, this concern becomes a substantial barrier to engaging in care. Duane and I agree that patients (or clients as you prefer) should neither put their therapists on a pedestal, nor see us as "wizards" (in any sense of the word). Licensed, professional healers have expertise to lend to the process of growth, but the individual who seeks to grow must remain in the driver's seat.

For many years, along with a deeply trusted veteran peer, I co-hosted a "new patient briefing group" when I worked at the VA. The goal of the group was to create the conditions for full engagement and successful outcomes in treatment. To help new patients see that they hold the power and the ultimate responsibility to make changes in their lives, I told them this:

> *Let's say that I was a body building coach – not a psychologist – and you came to me and told me that you wanted to compete in a professional body building contest. So, I gave you recipes to make, probably with lots of raw eggs and muscle milk in them – and gave you a workout regimen. You showed up week after week and told me that you were frustrated with your progress. I asked if you had done the workouts or made the lifestyle changes between our sessions and you said, "well no." In the same way, in therapy, we might reach some insights, and put our heads together to come up with a plan for change, but you need to invest the time and energy between sessions to bring about the changes you want for yourself."*

This simple analogy helped them understand their responsibility in therapy, which both empowered them and helped them to see therapists in a different light, as strategic advisors to the growth process, not saviors.

Military in the Rear-View Mirror is a thoughtful series of reflections on concepts that are critical for us to better understand – insights that can help veterans grow in positive directions. At times, it's hilarious (For example, as Duane recalls, "When I was in the Army, I literally had meetings to plan for meetings that would plan for future meetings. It's like staring down the hall of mirrors.") The overall tone of Duane's writing is realistic and challenging in the right ways. For instance, he says, "We didn't always love what we did in the military. We sucked it up then, we can suck it up now." This runs interference against the hindsight "glow" of "life in the military."

How many of us, in a generally adaptive way, screen out the negatives and bring a rosy glow to a previous period in our lives? Perhaps this is human nature – many of us idealize our "glory days," whether this golden period occurred during high school, college, or some other phase of life. It's even universally observed that many mothers and fathers recall childbirth and initial the transition to parenthood in glowing terms, even though this transition may have included terrifying and painful components, as well as peak life experiences. The implication in Duane's writing, echoed in my own writing on military transition, is that life is full of challenges. At many points in our journey, we need to dig deep and tolerate things that are difficult if we want to create a life that is aligned with our deeper purpose.

In his characteristically thoughtful way, Duane evaluates concepts with nuanced thinking. For example, in one passage, he takes us through an analysis of the benefits and costs of the "mission focused mindset." There are clearly several benefits to this mindset – among them, a "mission focused mindset" helps us set and achieve goals, helps others see us as trustworthy and reliable, and confers a feeling of pride in a job well done. However, there are some serious drawbacks to over-applying a "mission focused mindset." These may include "black and white/all or none thinking," and the tendency to foreclose options outside the range of what we initially perceive as aligned with the goals of our mission. In other words, if we keep our eyes rigidly fixed on one objective, we may miss out on other more valuable experiences and alternative pathways that may move us in a positive direction. This analysis is one of several in the book that shows Duane's thoughtful approach.

To wrap up this preamble, one of the greatest gifts we can give our warriors is to operate from the understanding that they can heal and grow when they are empowered with the right insights and resourced with the right team of support. In this vein, *Military in the Rear-View Mirror* holds important insights for service members, veterans, and those who care for them (whether as professional healers or those in other supportive roles).

"Doc" Shauna Springer, Ph.D.

Best-selling Author of *Beyond the Military: A Leader's Handbook for Warrior Reintegration* and co-host of Military Times' "Seeking the Military Suicide Solution" Podcast

Introduction: Why Veteran Mental Health

I talk a lot about changing the way we think about veteran mental health. When I do, I get a variety of responses. These include stigma, "Veterans just don't want to talk about mental health. They think it makes them weak." They include dismissal, "Veterans don't need therapy, they just need to get on with their lives." Sometimes, I hear, "but why Veteran mental health? Why not mental health in general? You know that's a problem for everyone." When talking about veteran suicide, others tell me that I shouldn't just try to help veterans, but help everyone.

Well, I can tell you, I don't know everyone, but I do know veterans. I may have a lot of tattoos on my chest, but none of them are a shield with an S on it; I'm not Superman. When I was in the military, I would tell my Soldiers: I'm not trying to change the Army, I'm just trying to clean up my corner of it.

Much of what you'll read in this book can and does apply to mental health in general. Our capacity for stress. The need for awareness. The methods used to intervene when a veteran is considering taking their own life are the same methods that we should use for anybody. The difference is that veterans experienced a lifestyle that is increasingly unique in modern society, and immersion in that lifestyle has come with certain psychological advantages and challenges that are also unique.

Solutions for Veteran Mental Health Can Be Applied Elsewhere

The fact that the military is a microcosm of society means that solutions that work for the veteran population could work for the population at large. This goes for employment, homelessness, and yes, mental health. The stigma against seeking mental health treatment is strong in veterans. If we figure out how to overcome the stigma in our nation's service members, then perhaps that message and method can be applied in other areas. First responders. At-risk youth. Domestic violence survivors. Yes, each of these are important and we should address them all; if we wait for one solution to apply to all of them, however, we will be waiting a very long time.

The Message Must Be Appropriate for the Audience

In my observation, universal solutions simply don't make the impact that they once had. Our society has become so individualized that we can tailor our education, our employment, and our consumer behavior to fit our needs and desires. The concept of market segmentation[1], in which we are grouped according common needs and interests, leads to messages that are increasingly tailored to a particular audience. The way we communicate today lends itself to this segmentation. If I wanted to reach an audience of military spouses that have children between the ages of eight and eleven within a radius of a certain zip code, I could do that.

The days of a single message for all people are long gone. We need a message, and a way of spreading that message, that is specific to the audience. While the interventions that are effective for veterans are also effective for at-risk youth, the way that we present those interventions are not. The ability to adapt our message to an audience is necessary when we are trying to sell books or shoes; it's also necessary when we're trying to sell an idea like utilization of mental health services.

The Mental Health Needs for Veterans are Unique

No matter which way we look at it, this is true. Certainly, trauma is trauma. Any event that threatens our lives or sexual violence to us or a close loved one is traumatic. Traumatic Brain Injury is not isolated only to the military. Colorado has become a leader in brain injury research; this is due to the frequency of head injuries as part of skiing and snowboarding accidents. What is unique is not the impact of these events; it's the source. Someone who experiences traumatic stress reaction from a vehicle accident, or a natural disaster is needs support, certainly. That person is not also having to contend with moral injury[2]. Or having to figure out how to meet their needs[3] after leaving the military. The repeated stress of multiple deployments and the strain that it has on the family[4].

This, of course, is also true for other populations. The system that supports at-risk youth can be as challenging to navigate for them as the Department of Veterans Affairs is for veterans. The legal system for domestic abuse survivors or the support system for victims of human trafficking have their own challenges. The argument is not that we should focus on veterans exclusively; it's that we should tailor our message and method of delivery to each population, according to their needs.

The Current Message is Not as Effective as it Could Be

Conduct an online search for "veteran mental health" and you will likely get a majority of responses that has to do with Posttraumatic Stress Disorder. There is an increasing awareness that veteran mental health goes beyond just PTSD and TBI, and many in the mental health community are talking about it[5]. The challenge is that the non-clinical aspects of veteran mental health, such as meaning and purpose, moral injury, and needs fulfillment, are not as prevalent in the conversation. If we are going to talk about veteran mental health, then we need to talk about *all* aspects of it, not just PTSD or TBI. Focusing on these areas alone leads to a less than effective message.

I am all for changing the way that we think about mental health in our nation. I absolutely support anyone trying to overcome the stigma associated with mental health treatment. Like I used to tell my troops, though, I'm not trying to change the world; I'm just trying to clean up my corner of it.

Thanks for taking the time to join me in that effort.

Duane K. L. France, MA, MBA, LPC
Sergeant First Class, U.S. Army, Retired

PART 1

Honest Discussions About Veteran Suicide

The impact of suicide in the veteran population is widely known. Many of us who have served have "a number;" the number of fellow service members that we have lost to suicide. For many of us, even most of us, that number is greater than the number of those we lost in combat.

There have been, and needs to continue to be, entire books written about how to end suicide in the military population. Not enough is being talked about it, because it's still happening. Until suicidal self-injury no longer becomes an option, more must be done. One of the first things is to have honest discussions about the topic; it's a taboo subject, one that we don't like to think about. There are tons of myths surrounding it and people doing things that we think are effective, but they're really not.

If you're reading this, it's likely that you have personal experience with suicide. It may have been a long time ago, and it may be recent. Even if we think we know what to do, it can be difficult to intervene when the time comes. It can be frustrating if we want to help, and we're not sure how; but the key is to try. And to keep trying.

The following short section includes a couple of thoughts on the topic. We need to learn how to have real and honest conversations about something that most don't like to talk about, but is a very real and persistent danger in the veteran community.

A Serious Look at a Serious Subject: Veteran Suicide

There is an epidemic of veteran suicide in our nation, and around the world. As has been identified by several different studies, the rate of veteran suicide compared to those who have never served in the military is significant. There are a lot of discussions about it, attempts to overcome it, and studies to understand it. Unfortunately, however, it continues to happen.

The significant challenge, however, is that veteran suicide is not the problem we should be focusing on. Yes, of course, it is a huge problem, and one that has personally impacted my life. I have had family members consider and contemplate taking their own life, and I have lost more of my former battle buddies to suicide since returning from combat than I did in actual combat. So yes, while it is a huge problem to be solved, it's not the main one that, in my opinion, we should be focusing on. We, as a nation, as a community, need to solve the underlying problem that leads to each individual veteran suicide. Identifying that problem is a challenge in and of itself, of course.

A veteran does not commit suicide just because they have PTSD. Studies have shown that a service member who takes their own life doesn't do so because they have been in combat; a 2014 study published in the Journal of the American Medical Association[1] did not see combat deployments as a risk factor, but instead a service member is at greater risk if they were 1) Male, 2) had a substance use disorder, and 3) had a preexisting mental health condition. And the greatest number of veterans who took their own life in 2014, according to a VA study, were ages 50 and older...cold war veterans, and perhaps gulf war and Vietnam veterans, but those who had not experienced combat recently.

In a series of episodes on the Head Space and Timing podcast, we looked at many of the different aspects of veteran mental health beyond just PTSD and TBI. Veteran suicide is a very real danger if any or all of these problems are not addressed. Besides just PTSD and TBI, a veteran with a significant substance use disorder may get to the point where taking their own life is an option for them. If they are struggling with depression or anxiety, it may have nothing to do with PTSD, but still dangerous. A lack of purpose and meaning, hopelessness about the future, is a significant contributor to the negative thoughts that surround veteran suicide. Moral injury, with it's impact of guilt and shame around things that were done or seen, can lead to struggle and inner turmoil. Not being able to meet our needs after the service...whether economically, stability in housing or employment...can lead to hopelessness and despair, or a lack of stable and

13

fulfilling relationships...any one of these can take a veteran to the brink of taking their own life.

Suicide is simply an extreme manifestation of a veteran's inability to manage these challenges in their life. The problem, then, becomes not "how do we keep veterans from taking their own life" to "how do we help veterans become aware of the challenges that they are experiencing, and help them reduce the impact of these challenges." Then suicide takes care of itself.

Suicide Happens on a Continuum

The thought of someone taking their own life is not an on or off switch, it's not that you're either suicidal or you're not. This is a significant misconception that a lot of people have. As you go from one end to the other, however, the danger increases. On the low end, a veteran can have some very vague thoughts, things like, "maybe it would be better if I just weren't here anymore" or "sometimes this sucks so bad, I can't deal with it." This is the first sign that our thoughts are heading down a dangerous path. Beyond these vague thoughts on the continuum are more specific thoughts: "I should just kill myself. Then all of this would go away." Again, heightened danger. If these thoughts are coming in and coming out of your mind, just in moments of stress, then something's up and you need to start talking to someone about.

The problem is, sometimes, these thoughts progress beyond just coming in and coming out of your mind, and a veteran can start to dwell on them. Think about them more often, and more intensely. Moving along the continuum, a veteran might actually start to move beyond the *if* they could take their own life, to *how* they could do it. A plan starts to come into focus, consideration of the method of them taking their own life. After this, the danger intensifies as preparations are being made: acquiring the method to put the plan in action, writing a note, making plans for after they are gone. Beyond making the plan and preparation is actually the attempt to take one's own life, and beyond that death by suicide.

So, here's the continuum again: vague thoughts, specific thoughts, dwelling on these thoughts, contemplating a plan, making a plan, preparing to carry out the plan, carrying out the plan, and death by suicide. The heartbreaking problem is, intervention is possible at any and every step in this continuum, all the way up to the final one. It is literally not too late until it is too late.

Some of the challenge, though, is that a veteran, or anyone considering taking their own life, can progress through these steps at their own pace and at their own time. It can be a long, slow-burning fuse, in which a veteran will remain in a stage of constant consideration or rumination for weeks, months, or even years…and then progress through the final stages in a matter of minutes. Or, it can be a short, quick-burning fuse, in which someone can progress from vague thoughts to putting a plan into action in the space of an hour.

There could also be many signs, or no signs at all. This is the danger in suicide, that it has the potential to shock and surprise those who are left behind, and along with the shock and surprise comes guilt, pain, and grief. The one individual who knows exactly what is going on is also the one who might have the least ability to keep it from happening: the veteran in crisis themselves. This is where honest discussions of safety come in, that there is a network of trust in place so that the veteran can reach out to a trusted, nonjudgmental source of support.

The Resources are Out There

I could talk about veteran suicide all day. Not because I want to, but because we have to. We have to have these honest discussions around the topic of suicide. If you're looking for more resources, you can find them here[2]. They include infographics that I've created that highlight the key points of some of the research that is done, or posts talking about the complicated nature of suicide. They include an impactful blog post and video that I put together to raise awareness about veteran suicide, We Lost Another Veteran Yesterday. You can see that here:

A colleague of mine, Tony Williams[3], is a fellow veteran and mental health professional. He once said something that has stuck with me: suicide is a national problem with a local solution. And that's the truth: we can make speeches about it, and write novels about it, but nothing will happen unless it happens in our household, our neighborhoods, our community. You can also listen to blogs on the topic, like Stacy Fredenthal's www.speakingofsuicide.com[4]. RENUMBER NOTES TO REMOVE #5

And finally, I want all veterans to hear this very clearly: the life you save tomorrow can and should be your own. The only veteran I can make absolutely sure will see the sun come up tomorrow is me. The only veteran whose life you can save tomorrow is yours. I desperately want to save every veteran's life, and I know that I can't. The responsibility is on each of us to have a real and frank conversation with those that love us, to make sure that we know: we're not alone.

The Boots Left Behind

There they sit, the boots left behind. Never to be laced again, not by the one who wore them. They're gone forever, but the reminders of them remain; the boots, to be sure, but the pictures, too. The memories. A small child in a fading photograph, grinning goofy smile. An awkward teen, nervously smiling in a pre-prom photo. A young service member, looking at the camera with a stoic, determined look; forever young. Forever gone.

You've heard it over and over again. A permanent solution to a temporary problem. A selfish act. A sin, even. All phrases to help someone to understand the impact of suicide. And some of the insidious problem is, in the midst of the pain and anguish that accompanies suicide, none of those matter. But perhaps this image of the objects left behind may make someone pause.

When I left home to join the Army, I left stuff behind at my parent's houses. Some of the stuff they kept, some they didn't. But what they kept, they kept for a really, really long time. The things we own, the things we have, remain here after we're gone. My grandmother: an amazing artist. Painting, sculpture, dollmaking, you name it, she did it. Many in our family feel grateful that we have some of these things, gifts from the past that remain after she passed away over twenty-five years ago.

When people die, their possessions remain. These objects have no meaning except what we give to them; but the meaning we give to them is significant. Sometimes bitter, sometimes sweet, sometimes both.

My father passed away a year ago; my brother, sister, and I have his things. He didn't commit suicide, but he's still gone. I have his keys, hanging on a hook in my office. I don't know what the keys go to. Whatever lock they fit will remain locked forever; or, if unlocked, will never be locked again. These are the everyday items that remain after someone leaves, and we hold them, or discard them. And the decision of what to do with them is not our problem when we are no longer here...that burden belongs to someone else.

This is not an attempt to guilt someone into not taking their own life. I recognize that those who are in a place where suicide is a very real possibility don't need any more negativity; they feel bad enough already[6]. Instead, this is meant to simply provide more facts to consider. More factors in the equation that I personally hope results in a veteran choosing to live rather than death.

What We Leave Behind Carries the Weight of Our Life

Think of the objects that are meaningful to you. A piece of jewelry, perhaps, or a favorite book. The chair we sat in more than all the others, the things that you look at and think, "mine." Then, think of the objects you own that are not meaningful to you; that half-used bar of soap. Those old shoelaces. Those boots that I mentioned at the beginning. All of these things remain after we leave, and must be kept or not kept. After my father passed away, the possessions that we decided to keep filled half of my garage for six months. They were constantly there, a reminder of his passing, until I got around to doing something about it.

What We Leave Behind is a Reminder of Our Death

In some cases, the objects that remain are treasured items that remind us of the loved and lost, and are cherished. My grandmother's paintings. My father's keys. These things we pick up as we travel through life and leave behind us can be appreciated as legacies. The things we leave behind after we cut our own lives short, however, bring a different reminder. A reminder of what might have been. For some, it is a reminder of what was or was not done. They are reminders of a question: did I do enough? Was there more that I could do?

What We Leave Behind is a Reminder of a Future That Will Never Be

If suicide is anything, it is an end of tomorrows. For someone in pain, that may be enticing, and entirely the point; but if there is an end of tomorrows, there is an end of everything that tomorrow brings. More pain, yes, because that is the human condition, but also the joys. Generations that come behind. Satisfaction of a life well lived. All of the interactions that are part of the human condition: graduations, and celebrations, and recognitions of effort.

In Ulysses, Alfred, Lord Tennyson, says: "Old age hath yet his honour and his toil." There is respect for the old soldier, even if the old soldier feels alone and abandoned; and there is work to be done, either the daily work of family and friends, or monumental work that continues to impact the world. Let us reach those days, and not cut them short; let us continue to strive for both the honor and the toil

PART 2

Who We Were: The Past Impacting the Present

One of the most significant parts of post-military life is an identity as a former service member. There are those of us who someone sees and they can immediately identify as a veteran: we wear the hat, the clothes, we have the gear, we keep the stuff on our desk or hanging on our wall. If you walk into my office when I'm no there, you will look around and be certain that this person served in the military.

There are others of us who you think may have served, but aren't sure. Maybe it's the way we talk, or carry ourselves. Our demeanor. These snap judgements aren't always accurate; often, people think that I may be a high school wrestling coach or a local college football team's D-line coach. I didn't wrestle or play football in high school, but I guess I have that look about me. There is something that others pick up about a veteran, though, especially other veterans.

Then there are those of us who you would have no clue that we had ever served. Consider famous actors or industry leaders who served in the military; it had become part of their past, maybe an important part, but it informs who they are today. It doesn't still have a current day-to-day impact.

There's nothing wrong with any of these styles. As I often tell my clients, if you are happy with the way you are, and you're not hurting yourself, your family, violating someone else's rights, or doing anything illegal, then go for it. It's not up to me or anyone else to judge someone else.

The difficulty comes when our past impacts our present in a negative way. A current or former service member is psychologically impacted by the military, in the same way that a teacher or a construction worker is psychologically impacted by their career choice. It's a hard job, like many, and it's dangerous; that changes people. This section looks at how the past impacts the present, and what action can be taken after awareness.

The Once and Future Warrior

There are many obstacles for a veteran seeking mental health counseling. One is that veterans believe that they have to admit that they are weak or broken before they get up the courage to go talk to a therapist. That they are somehow deficient, less than what they were, or lacking in some critical way. In my experience, mostly everyone wants to have a positive image of themselves. Service members see themselves as capable, confident, able to get the job done. The challenge, though, is that the veteran sometimes forgets where they came from[1]...they seem to forget that they are a warrior.

When I see a veteran for the first time, I don't see them as broken. I don't see them as weak or in any other way deficient. I see them as someone who, at one point in their life, was able to overcome significant obstacles. To thrive under considerable stress. To do something that few others have done or are willing to do. Raising your right hand to volunteer isn't enough, of course. Anyone can do that. I can tell you, the first time I did that in 1992, I had very little idea what it would be like. Of course, once I got to basic training, I found out pretty quick.

Deploying Does Not Make You a Warrior, Overcoming Adversity Does

The word "warrior" was not really in the Army's vocabulary for much of my military career. This may be hard to imagine for many of the younger veterans, but the Soldier's Creed[2] (and the Warrior Ethos that it contains) was not developed until 2003, after the nation had been at war for over a year. Before that, us crusty non-coms used to refer to our soldiers as "troops" or "hero" or any other of a number of names. But the idea of Warrior was not common. Sure, we screamed about "blood, red, red blood" during bayonet training in boot camp, but Warrior? Not really.

Reflecting on the concept of warrior, though, I consider someone who has been battle-tested. Laying your life on the line, literally, and coming out the other side makes one a warrior. It can also be argued, however, that anyone that has overcome adversity of a significant measure can be described as a warrior. More than just the stressful boot camp, but actually being tested and found to be capable. Any adversity...a fire on the ship, an ambush, an improvised explosive device. Anything that leaves a scar, physically or psychologically, is adversity, and overcoming adversity and persevering makes one a warrior.

Veterans Were Once Warriors

Looking at things from that definition, veterans are warriors. It doesn't matter if the adversity was a running firefight through the alleys of Baqubah, or manning aircraft that escorted Russian jets out of NATO airspace during Cold War incursions. The military is an inherently dangerous occupation, and the stories that veterans have of the numerous adversities that we've faced would shock and surprise many who haven't served.

The vast majority of veterans I meet have overcome significant hardships in their life. Sometimes, the hardship began before they joined the military, and life in the service was a cakewalk compared to their childhood. The hardship was deployment, danger, combat, war. At some point in their military career, veterans overcame their own doubts, fears, and hesitation and conducted themselves like a warrior. Why they did it wasn't important; it might have been to get the glory, or impress the guys or gals back home, or to save their buddy and bring them back home. The reason is not important, but the result is: the scars remain. The proof of the Warrior, whether the warrior wants to admit it or not.

Adversity Impacts the Warrior

The problem often is, the adversity changes us in ways that we did not expect, and don't realize until after the change has occurred. Adversity, trauma, stress, combat, any of these things...veterans are no longer the way they were before they experienced them. "Older and Wiser," some might say, but no longer joyful and careless. We know the darkness that lies in the hearts of our fellow man, and what lies in our own hearts...often because we've had to engage that darkness in order to overcome adversity. Scars aren't handed out like candy; whether they are psychological or physical, they are earned, through pain and stress, and are only healed with time and treatment.

Therefore, the one thing that makes one a Warrior...adversity...also has a high likelihood of changing the service member in a critical way. Sometimes in a detrimental way. That doesn't mean that the warrior is screwed up...but the situation certainly was. We will fight to get treatment for things in the environment that caused us physical illness, such as Agent Orange, Gulf War Syndrome, or toxic burn pits...and we recognize that it was the environment that caused this, not some personal weakness or deficiency. When it comes to the psychological impact of adversity, however...that's taken as an entirely personal weakness, and the environment had nothing to do with it.

Veterans Can Be Warriors Again

The key about being a warrior...once you have come out the other side of the fire, you can do so again. You don't gain strength by avoiding adversity, you gain strength by overcoming it. The person who has never lost, has never bled, has never failed has never been truly tested…and is, in reality, no warrior at all. The scars of a warrior remind us of three things: we are human, we have persevered, and we can heal. When we find ourselves at a point where we feel defeated, we can and should remember: we were a warrior once, and we can be a warrior again.

Why Would Veterans Want to Go Back to Combat?

If you had a choice, after all these years, would you go back? Absolutely. In a heartbeat. This is something I hear from veterans I work with all the time: if they called, I would go back in an instant. This may come as a surprise to those who have never served, and I certainly don't mean to say that all veterans feel this way. But for many combat veterans, there is an often conflicting desire: to remain home with their loved ones or to go back to war. This has elements of the fact that it was often the best time and the worst time[3] in our lives. And it also has to do with the back there paradox[4] that some veterans need to resolve.

This isn't something that is just reserved for current veterans, either. I have an uncle who was a Vietnam Veteran, and when I deployed to Iraq, he said, "Take me with you, kid! I can get some as well now as I could then!" Forty years after he had been in combat, he wanted to go back. In some ways, almost *needed* to go back.

But why is it like this? If we really hate war, why would we do it again? It's certainly not for the glory, because there isn't a lot of that. It's not for the money, either, because nobody gets rich in the Army. There reasons that a service member may want to return to combat are more personal, and other-focused. It doesn't mean we don't love our families, it doesn't mean we have a death wish; it has to do with the values that we developed when we were in the service. Here are some of the reasons that a veteran may want to go back to war:

A Place of Skill and Familiarity

For a service member, a deployment is the end-all and be-all of the reason we joined in the first place. Maybe you said you joined to earn money for college…that's certainly the reason I gave myself…but it very quickly developed into a love for the military, the camaraderie, the enjoyment of it. It goes to the paradox where we hate war, but love combat[5]. Deployments, and especially combat deployments, were a place where we knew how to use the skills we had developed. What we did was significant, meaningful work. Life could be very simple while deployed; get up, do your job, get back to base, and hit the gym/mwr/chow hall, whatever your routine was. Sure, it got boring and repetitive, but we knew how to do stuff and we knew where things were. It was familiar.

To Preserve the Meaning of Sacrifice

Another reason I hear for wanting to go back is simply the sense of unfinished business. I was with a group of Marines when we were watching news coverage of the fall of Fallujah[6] in 2014. There was significant anger;

one of them turned to me, and said, "Why the @#$% did we go through all of that, then?" The fact is that many veterans who returned home feel as though the true heroes were the ones that didn't return. There is a personal desire to ensure that they did not give their lives in vain; to ensure that there sacrifice still has meaning. Of course, it has meaning to us personally, and if we tell ourselves that it no longer has meaning, then it doesn't to us. We have control over the meaning that we give to things. This may not be a particularly rational or practical reason, but it's still a very powerful reason nonetheless.

Our Brothers and Sisters

These first two meanings are very personal, although they are tied to the next two. These reasons, however, have much more to do with others. It's been said often in many places: the reason the grunt on the ground fights is not for the country, or the political reasons, but for the people to their left and their right, in front and behind. Because of their fellow service members. The memories of our brothers and sisters are usually the heaviest and most long-lasting: I made it back. They didn't. I made it back because they saved my life. We were a team, had a bond, did something important together. Many veterans would go back to combat if their brothers and sisters they served with called and said, "hey, we need you." Many would be there. In a heartbeat.

So That Others Don't Have To

This is a reason for wanting to return to combat that exists on a much more theoretical scale. We sacrifice so that others don't have to. This was always a driving force for me; I served in order to provide a life for my children in which they did not have to. Of course, it's challenging when the ones who we sacrificed for don't recognize that sacrifice; an incident in which a teenager vandalized a memorial[7] honoring Medal of Honor recipient Michael Murphy brings the lack of awareness of and appreciation for that sacrifice into focus. Still, the one who serves doesn't always do it with the expectation that the beneficiaries of that service will appreciate it; they will continue to do so. It's ingrained in many service members; we carry the heavy load so that others don't have to, knowing that they would do the same for us. We go the extra mile, because we know that we won't be walking alone.

It can be challenging to consider whether or not we would go back. It's not just nostalgia, the desire to return to the days of our youth; our reasons for wanting to return are based on our values and beliefs. And, again, it doesn't mean we have a death wish or don't care for our families; it means that the words honor, duty, and sacrifice are more than just words to us, and those words don't lose their meaning when we leave the military.

Fear and the Fading of a Hardcore Soldier

A friend of mine, Special Operations Soldier, once told me: "It's one thing to think that fear is weak, that suicide is the coward's way out. I thought the same way, once. It's totally different when it happens to you, and you *know* that you're not a coward."

When a service member or veteran starts to struggle, either when they're in the military or long after they get out, there is a significant element of doubt and shame. The military is such a team-focused culture that, when you feel like you're no longer part of the team, then you feel cut out. Isolated. Often, this isolation is self-imposed; we beat ourselves up worse than anyone else could possibly do it.

Fear is what you strike in the heart of *other* people…it is never something that is found in *us*. The challenge, of course, is that fear is a natural emotion that occurs in everyone. No matter how much we want to deny it, fear is there; many times, for a very good reason. Much of the problem with fear comes from the fact that many service members believe that the presence of fear makes them less than what they were, and the hardcore person they were starts to fade in the light of the person they see in the mirror. The person the service member sees in the mirror is inferior. Weak. Pitiful. If the person we see is a faded shadow of who we used to be, then we start to hate ourselves. And that's when the problems start.

Service Members are Dependable…

One of the things that many service members and veterans pride themselves on is their dependability. They could be counted on…by their buddies, by their unit, by their nation. They look down on others in the military who don't pull their weight, who can't be counted on. The military is a talent-sucking organization; it disproportionately relies on those who drive harder, move faster, work longer and harder than others. They will be asked to drive even harder, move even faster, and work even longer and harder…and they do. It's not because the unit leaders are sadistic. They, too, are often driven to move faster, work longer and harder by their superiors. And so it happens all the way up the line. The dependable rock of the unit is the one that everything is built on. Being that rock is a point of pride for many.

…Until They're Not

Once, in Afghanistan, a soldier was coming off of an eight hour guard shift. Another element of his unit was out in sector, seven hours away; a unit leader told him and another soldier to get ready to go out. The soldier resisted, claiming safety and crew rest. The leader chewed him a new one,

24

and even made sure to make sure the guy lost rank…what the leader didn't know at the time, because the soldier didn't say anything to anyone, was that he was struggling with a painkiller addiction at the time. He had just taken some painkillers, and knew that if he went out into sector, he would be putting more lives in danger…and the shame of it all crushed him.

The soldier was in pain, he was addicted, and he had just let his unit and his buddies down. He got busted in rank and gained a reputation of undependability…because of his secret. It wasn't until years later that the soldier found the leader and told him what happened and why it happened…and that he had gotten help.

This story has probably been repeated over and over again. And continue to happen, until things get under control. When fear, anxiety, depression, any of these things start to develop in a service member or veteran, they feel like they've let themselves down. Their buddies. Their nation. They feel their former dependability start to fade…and they become less of who they were and don't like who they are.

A Hard-Charging Service Member is Irreplaceable…

Another point of pride for many service members was the fact that they were the one who could be counted on. It's not an ego thing, "look at me" kind of stuff. That's what they've been told by their leadership. It's what they've been told by their buddies. "We can't let you go, you're the key." When the hard work needed to get done, when the mission needs to be accomplished, they were the once to get the call. And like the hardcore person they were, they got it done. The pride in a hard job well done is a significant motivator.

…Until They're Not

When I got ready to retire, I was the Operations NCO for a special operations support company. As the time for my retirement got closer, my First Sergeant said, "take some time to get things prepared." I looked at him in surprise…I said, "I'm not ready yet, I still have to make sure things are set before I go." He responded, "Yeah, but we'll have to figure it out sooner or later…we might as well try to figure it out now while you're still around. We'll give you a call if we need something."

And that was considered a *good* transition, a great leader taking care of his troops. I've heard many service members who were worked all the way up until the day they were discharged, then felt discarded. Thrown away. The song *Confusion* by Metallica[8] describes how many service members feel, especially after being injured or struggling with the psychological impacts of their service:

Label him a deadwood soldier now
Cast away and left to roam

This kind of experience causes the veteran to feel even more expendable. It's one thing to know that we're prepared to give our life for our country, but for us to sacrifice our self-worth, our identity? That's a hard load to carry.

And it doesn't have to be that way forever. There is a path back to a place of stability, where we can be just as hardcore in our post-military lives as we were in the military. In a different way, maybe, but still dependable. Still irreplaceable...to our family, to ourselves. And that's never a bad thing.

Veterans are Masters and Creators of Chaos

One thing about being a veteran; when we were in the military, we could put up with just about anything. Extreme boredom. High stress. We would stand out in the rain, for no apparent reason, just because someone said that's what we had to do. Our natural environment was chaos. And we performed extremely well under stress and in chaotic situations.

Thinking back on my own military career, I was really in the zone when I was managing a chaotic and rapidly moving situation. In training, usually at the Joint Readiness Training Center or the National Training Center, I'd go for 24 to 36 hours with no sleep. SOF guys did one better; Ranger school is two months of no sleep, no food, and constant chaos. And in combat? This is where it really gets to you. Days of intense boredom, the same day happening over and over again, and you find yourself wishing for something to happen. Really? Against all of our better judgement, not to say our own personal preservation instincts, we want someone to shoot at us? Hard to believe, but it's true.

Many veterans think, "If I'm not one foot ahead of disaster, then I don't feel alive; dancing on the razor's edge is where we feel most comfortable." We've all heard the stories of returning veterans finding themselves barreling down the highway at 90 miles per hour, or jumping out of airplanes, or going down to the bar just to get into fights. It's adrenaline, it's misplaced anger…and it's a desire to create an environment where we know how to manage ourselves: chaos.

We Don't Want Chaos, but We Cause It

This is a challenge when it comes to what we think we want and what we actually do. If you ask me, or many veterans, "When you leave the military, what do you want?" "Peace," they say, "A shot at the American Dream[9]. To just take care of myself and my family, to be happy, to be at rest." That's what we *say*…but what we do is totally different. We can't find that we are able to relax in non-military life. We're on edge all the time. Why? Do we want to? Most of the times, not really; we're just operating in an environment that no longer exists.

We're using old skills that used to work in one situation but don't work where we're at. We don't adapt; we're fixed in our old patterns of problem solving. By not adapting to the new non-military situation, we are creating chaos and opposition in our lives, which makes us uncomfortably comfortable.

We Create Problems in Order to Solve Problems

So, in order to feel like we're in control, we create problems where there weren't problems before. We perceive problems that don't exist. We're not doing this consciously or deliberately; without awareness, without self-reflection, our chaos creating ability is on wide display and in full force. If we don't stop and reflect…if we don't use the space between stimulus and response[10] that Viktor Frankl described…then the default setting is going to be Tasmanian Devil. Anything that happens…chaos. If anything *isn't* happening…chaos. We create chaos so that we can manage chaos, we create problems where none exist in order to solve them. I'm not talking about all veterans, of course, but I see this often enough in the veterans I work with that it happens more often than we might want to admit.

Some Veterans Feel They Deserve Chaos More Than Peace

Another challenging problem when it comes to chaos creation is that many veterans are carrying so much guilt from things that happened or didn't happen when they were in the military. This is as much moral injury[11] as it is PTSD or anything else; some veterans simply don't believe they deserve peace. They feel as though they're condemned, and condemned people get a choice of their last meal and not much else. The problem with thinking we don't deserve peace, don't deserve forgiveness, is that we don't seek it. We don't work to be comfortable with either of them. This is a belief that we have about ourselves as it relates to the world. As with many other beliefs, it becomes a self-fulfilling prophecy; and our lives are filled with chaos.

Chaos Happens in Small Ways, Too

I'm not immune to this, either. When it comes to free time, I don't have much of it, because I don't allow myself to. This isn't a good thing; my wife once said, "When you retired from the Army, I thought we were going to see *more* of you, not *less* of you." She was joking, but then again, not really. But I sometimes feel like I have a congenital birth defect that keeps me from saying *no* to things. This has happened in cycles since I've left the military; I find myself bored, so I look around for things to do, and very quickly find myself overwhelmed and over extended. So I pull back, then find myself bored, so I start looking for things to do…and the chaos cycle begins again.

Awareness and Balance are Key to Combating the Urge to Chaos

As always, awareness is the first thing that is needed to counteract our urge to create chaos where none exists. Without the awareness that we engage in this cycle, we can't disrupt the cycle. We may never be comfortable not doing anything, but we don't have to always be comfortable in chaos, either. Finding balance between calm and chaos is a way for veterans to find meaning in their post-military lives.

On Courage, Vulnerability, and Radical Honesty

"Vulnerability sounds like truth and feels like courage. Truth and courage aren't always comfortable, but they're never weakness."
— Brené Brown

Like all other branches of the military, the Army has a number of core values. For us, it creates the mnemonic LDRSHIP: Loyalty, Duty, Respect, Selfless Service, Honor, Integrity, and Personal Courage. A common question from leaders to their subordinates: "what is the most important Army Value, and why?" A good ol' standby for promotion and selection boards.

Before 9/11, my answer to that question was Integrity. Without integrity, you couldn't have the rest. On 9/11, though, I thought of those who sacrificed themselves so that others may live. The firefighters who were running up the stairs while everyone was running down. The story of Rick Rescorla[12], who had proven his courage many times over in his lifetime, and continued to do so until the end.

Of course, that's the courage we think of when we talk about the word. We don't often equate courage to vulnerability, but that's exactly what those 9/11 heroes were doing…putting themselves in a vulnerable position, the ultimate vulnerability. When we talk about personal courage, though we also need to think about emotional courage. The courage to be open and radically honest; the complete, total, and one hundred percent honesty about things that most just don't like to talk about. Which takes more courage: to hide that which we are ashamed about because of what we believe what others will think about us, or to step out in spite of that shame with no regard of the opinions of others?

I've had a number of conversations with veterans who were radically honest about themselves and what they experienced or are experiencing.

Radical Honesty About Our Actions

"Confront the dark parts of yourself, and work to banish them with illumination and forgiveness. Your willingness to wrestle with your demons will cause your angels to sing."
— August Wilson

There are things that we do in our lives that we're just not proud of. It's easy to keep them hidden, and for some, it's good to do so. Other times, however, keeping the secret locked away only makes it worse. Radical honesty about our actions is being open and vulnerable in admitting that the things that we were doing weren't helpful. In many ways, they were downright destructive. My friend and colleague Justine Evirs was radically

29

honest in our conversation; while she was exhibiting all the signs of outward success, making a difference in the lives of others, she was struggling with a significant alcohol addiction. During many of our conversations before and after the episode, and even during it, she expressed her anxiety about talking about it publicly. And that's understandable...but giving in to that anxiety isn't courage. It's capitulation.

Along the same lines, Marine Corps veteran and veteran mental health researcher Kate Hendricks Thomas[13] was open and honest about her reactions to combat. Her statement, "I realized that sitting in the basement drinking beer alone in the dark wasn't healthy" is a phase that has stuck with me, and one that I use often to describe how destructive our behaviors can be. Her article in Task and Purpose about how she hid her abusive marriage[14] is also an example of radical honesty. She says, "Marines are supposed to be tough." The toughness comes from being radically honest, not hiding in shame.

Radical Honesty About Our Mental Health

Courage is being scared to death ... and saddling up anyway. — *John Wayne*

That's what this book, and the other things I do, are all about: open and radically honest conversations about something that most just don't like to talk about. It's not perpetuating the myth of the broken warrior; all humans are flawed, we are all broken in our own individual way, and to deny it is to perpetuate a lie. There have been many stories of veterans being radically honest about their struggles with mental health; my buddy, Green Beret Jeff Adamec, is one of them. He said in our conversation, "It's one thing to think that mental health problems are cowardice and weakness; it's something totally different to experience them yourself, and know that you're not a coward." The most elite soldiers, the top of the pyramid when it comes to the profession of arms, are not immune to reactions to traumatic stress, any more than the rest of us are.

And being radically honest about our mental health is a universal trait. UK Veteran Andy Price was open enough about his struggles after leaving the military to be open about it. What's more, he decided to do something about it. If he is sitting in his world, feeling the way he does, then he knew that other veterans were hiding their own private shame. He started to talk about it, to show the world what's really going on inside of our heads. He literally shone a light on it by allowing his story[15] to be told publicly. If that's not courage, I don't know what is.

Radical Honesty About Our Lives

"To share your weakness is to make yourself vulnerable; to make yourself vulnerable is to show your strength." — Criss Jami

Sometimes, we need to be honest about our actions, our thoughts, and our emotions. Our entire lives. I met Adam Schumann after a talk that he and author David Finkel gave in Dallas, and Adam's life is a literal open book. The book and movie *Thank You for Your Service* is about Adam's life; being medically evacuated out of Iraq not for physical wounds, but for traumatic stress reaction. His experiences upon returning home. The challenges, triumphs, failures, and successes that go along with making his way in his post-military life. A willingness to be radically honest about everything, not so that it can benefit himself, so that perhaps it may help others suffer less.

That's the psychological and emotional equivalent of running up the stairs while everyone else is running down.

The Dangerous Trap of the Comparison Game

He could measure his own progress only in relationship to others, and his idea of excellence was to do something at least as well as all the men his own age who were doing the same thing even better. The fact that there were thousands of men his own age and older who had not even attained the rank of major enlivened him with foppish delight in his own remarkable worth; on the other hand, the fact that there were men of his own age and younger who were already generals contaminated him with an agonizing sense of failure – Joseph Heller, Catch-22

One of my favorite wartime novels, Catch-22. Consider the irony of it: reading a satirical account of military life in a war zone while actually in the war zone, and laughing so much that it made you cry because of how accurate it was. Joseph Heller[16] was a veteran, and he knew what he was talking about; the absurd characters in the book were not so absurd, looking at it from the veteran viewpoint. Some of them were hauntingly familiar and a little too close to home.

One of my favorite passages is the one quoted above. The character, a Colonel, was delighted that there were men older than him who had not even achieved his own rank, but enraged that there were men younger than him who had risen higher in rank. How often do we do this to ourselves? Not to this extreme, of course, but somehow compare ourselves to others and come up feeling like we are less than worthy.

This goes beyond just "keeping up with the Joneses" or needing to fill our lives with status symbols. It speaks to our own sense if inadequacy, our inability be confident in our abilities. Whatever I've done, others have done it better, faster, quicker. However far I go, others have gone farther, faster, quicker. And if this was true in the military, it might certainly be true in post military life. If you were in for eight years, you go back to college at 26 surrounded by eighteen year olds. You look at the bright young minds and feel old, and slow. If you retired, you enter the corporate workforce at 42 rather than 28; again surrounded by bright young minds. Here are some thoughts on how the comparison game can trip you up.

The Comparison Game for One

Sometimes, you might not be playing the comparison game as a team sport; it might be a solo venture. You might look at yourself now, and compare yourself to yourself twenty years ago. I was stronger, faster, quicker, lighter. I was in the 82nd Airborne Division when I was 25 years old, and it seemed like I jumped out of every aircraft I could board. We called it "jump chasing" and I was doing it to be able to qualify for senior Jumpmaster wings (funny what a man will do for a tiny bit of medal). And I loved it. My next Airborne assignment? Thirteen years later, as well as after two combat deployments and about twenty pounds heavier. Jumping out of airplanes (the Army way) is much different when you're 37 than when you're 25. When I was 25, I could jump two or three times a week; a couple of times, twice in one day. Thirteen years later, though, it took me a month (or two) to recover from a jump.

What's your solo comparison game? Getting out from underneath the shadow of your past? I'm not as good as I once was? As the Toby Keith song[17] goes, though, we can be content in being as good once as we ever were. We can recognize that now is now, and shut down the comparison game altogether.

The Comparison Game is a Sham

Let's face it, nobody has it all together. Everyone has challenges, dents and bruises in the fender. If I didn't know this after being inside my own head, I see it with the veterans I work with every day. So when we look at others and compare ourselves to them, about how they may seem to be more effective, efficient and successful, we don't know what's going on behind their eyes. We only see bits and pieces of their lives, the parts that they show the world, and of course those are the best parts. It's as if we're all backstage projecting live images of ourselves. And the online life has made it worse, in some respects.

It's been said that we compare our real lives to other people's highlight reels[18]. The fact is, however, online life is contrived. It's a curated display of what's going on behind the scenes. How's this for real: the reason I know the comparison game? I play it myself. I have to force myself to stop playing it, but I know what it's like because I do it. And when I question it, I know that it's a sham. Ain't no one else being the me I am right now, so there's no need to compare myself to them.

The Comparison Game Will Hold You Back

It can really drag you down. The power of the mind is amazing; we can do anything we put our minds to, but if our mind is the thing that's holding us back, it will really hold us back. The feelings of self-doubt, the imposter syndrome, the comparison game will de-motivate you. When we're de-motivated, we tend to do less. Pushing through those voices of doubt, refusing to play the comparison game, is the way to get moving again. You do you, because no one else is going to. No one else can.

And that's enough.

Arrogance and Humility, Entitlement and Invalidation - Post-Military Mindsets Holding You Back

In my experience…and this is as much personal experience as it is observation…veterans tend to underestimate their own abilities and capacities. Whether it's because the military is a collective culture where there is a set parameter of "acceptable" behavior, or because of the evaluative nature of the military, we are and can be extremely self-critical.

Maybe it's because many service members internalize the values that the military attempted to instill in us. Selflessness, sacrifice, individual effort is for the benefit of the whole rather than the individual. In the military, the concept of "my shield covers my brother and sister" is one that carries into other aspects of our lives. Praise is deflected; leaders point to the efforts of their subordinates, troops point to each other when asked which is the bravest, the worthiest. There are exceptions, of course, like there always are, but this is a character trait that I have seen in many veterans.

Self-confidence Becomes Arrogance

It's almost as if our evaluation of what is acceptable slips a couple of notches. If we measure arrogance, healthy self-confidence, and humility on a scale, arrogance would be a ten, self-confidence a five, and humility a zero. For many of us, the scale drops…arrogance becomes a six, so that if we express what is generally considered self-confidence, it is approaching arrogance. We don't want to brag, when all we're really doing is accurately describing what we have the ability to do. That means that what we see as self-confidence is really borderline humility, and what we consider humble is down the charts.

This could be a source of discomfort with being told "thank you for your service," especially in the younger generation of veterans. "Just doing my job" is a typical response that many have to that phrase.

Comparison and Competition

Another possible source for a shift in self-worth is the pervasive nature of comparison and competition in the military. Service members are very literally ranked in comparison to those around them. One glance at someone's uniform lets them know where they stand in the hierarchy of the military compared to someone else…and, often, that extends not just to people of different rank, but who has been one rank longer than someone else. Who has more "time in grade" is something that many people know, because that conveys a certain perceived level of competence, either good or bad.

Another aspect of the military is the inter-service and intra-service rivalries that abound. We cut each other down as much as any college football rivalry. For paratroopers, those who are not Airborne qualified are "dirty nasty legs." Service members who serve combat occupational specialties versus those who serve in support specialties. Those in support specialties who serve with combat troops, versus those who didn't. Those who have deployed to combat, and those who haven't. It's as if the entire military experience is measuring where I stand in relation to others, and it's hard not to internalize that sense of comparison.

Explanatory Style

Explanatory or Attributional Style is a way that we describe the meaning of events to ourselves. It's how we explain what happens, and we can apply both positive and negative explanatory style. The concept of explanatory style emerged from the work of Martin Seligman and helplessness theory[19]. There are three domains of explanatory style: personal, permanent, or pervasive. Personal explanatory style is the length to which someone describes an event as having an internal or external cause; if something bad happens, it was either bad luck or my fault. If something good happens, it was either my skill or a fluke. For some veterans, we apply a negative personal attribution to positive events...we dismiss it as due to external factors, not our own ability.

Consider this scenario: a service member leaves the military. College is paid for, so they go back to school, but I'm not really sure if I can do it. High school was so many years ago, I probably forgot everything, right? Besides, school ain't for me. In spite of that, I doubtfully move forward, anticipating failure and expecting the worst. But...it actually seems easier than I thought. Can this be true? Nah, I must have just lucked out and got an easy class. Well, I'll keep going...and end up on a roll. Dean's list. 3.8 GPA. Something in the back of my mind, however, is certain that the next class is going to be the one that kills me, imposter syndrome is going to kick in and someone is going to bust through the door and shout, "what are you doing here?" Then everyone is going to know I'm a phony.

When Not Enough becomes Too Much

As I mentioned before, this doesn't apply to all veterans. We are not all one single homogenous group that thinks and talks the same. We all know people whose arrogance scale has slipped the other way, those that seem to believe that their military service entitles them to a higher status than those who never served. That attitude is as unhelpful as the other type of attitude..."me, me me" all the time is as imbalanced as "them, them, them" all the time. We have to be careful not to get TOO full of ourselves, just as we have to be careful not to sell ourselves too short. Balance is key.

Letting Go of Unfinished Business

What has been done is enough. It's hard to think that, for many veterans. Sometimes, those who served struggle with the thought that they could have done more. Whether you served three years or twenty-three, there is a feeling of leaving something on the table, leaving a job undone, somehow failing because not enough was accomplished that leaves veterans unsettled.

Sometimes, that feeling of unfinished business can go deeper. This is especially true when it comes to the loss of a brother or sister; "what more could I have done?" "What if…" When the stakes are extremely high, the pain of unfinished business can be immense, and sometimes unbearable.

It's those "what ifs" that are gonna get ya.

The challenge, though, that we often feel the weight of unfinished business when the stakes are not extremely high. By dwelling on unfinished business, we're subconsciously giving ourselves permission to pass judgment on our actions. As service members, we prided ourselves on a job well done. We expect others to do a job well. When a task was to be performed in the military, we were expected to do it, or expected it to be done. We measured others by their ability to complete the tasks set before them; so, of course, we measure ourselves in the same way. Perhaps even more so.

Of course, this is true outside the military as well. You're not going to get very far in any job if you leave things half done or incomplete. The challenge arises when we take things too far; when we don't give ourselves credit for the work that has been done, and call it complete.

What Was Done Was Done

I'm not a big fan of the phrase, "it is what it is," but it's a statement of basic reality. What happened, happened. There's no denying that the thing occurred. Whatever we did in the military has happened. The good, the bad, everything that happened in the past is as complete as it's ever going to be. We can have regrets about what we didn't do, things that we didn't finish, things that others did and we didn't. To what end? What good can come of it? Sure, we can learn from mistakes and make sure they don't happen again, but I'm not going to be leading a platoon in Afghanistan again unless something goes really, really wrong in the world.

What Was Done Was Enough

Did you serve six years in the Army Reserve in the mid-1990s? Three years in the Marine Corps in the early 1980s? What you did was enough.

You served your country; nothing more was asked of you. When service members raise their right hands and enter the military, we don't commit to deploying to combat or jumping out of airplanes or doing a thousand of things that are possible in the military. It only requires us to defend against enemies, bear true faith and allegiance, and obey orders. If, after our time in the military is done, we can say that we did those things…finished business.

The problems begin when we put more requirements on ourselves. When we fall into the dangerous trap of the comparison game[20]. We look at what others have done, measure ourselves against it, and call what we did incomplete. Unfinished. Unworthy. Not only do we not give ourselves credit for what we did do, we take value away from it, and make it seem worse than if we had never started it in the first place.

What Wasn't Done Wasn't Necessary

Perfectionism, like the "what if" game, is going to get you. "Don't let the perfect be the enemy of the good," as the phrase goes. The term "good enough" is like fingernails on the chalkboard for me, and for many veterans, but often, it's true. History is filled with examples of people tinkering with things far beyond what needed to be done. Steve Jobs and Apple weren't the first to include graphic user interface or things like a mouse in personal computing; that was done by Xerox in the Palo Alto Research Company[21] in the late 1970s. It took someone like Jobs, though, to walk into the PARC workshops and see what was being endlessly developed and say, "what you have here is good enough" and ran with it.

One deployment? Two deployments? It wasn't necessary for you to do more. We may tell ourselves that it was…that doesn't make it true. What was left undone didn't bring the universe to an end. If we believe what we tell ourselves…we didn't do enough…then why wouldn't be believe it if we tell ourselves something different? We *did* do enough?

What Wasn't Done Wasn't Yours to Do

There's one more aspect of unfinished business. Sometimes, the thing that was "left undone," in your mind, was through no fault of your own. I have worked with veterans who feel extremely guilty because they went on one deployment with their unit, and then changed duty stations. Any service member or military family member knows that when you're told to move, you move…see the Oath of Enlistment again. At the next duty station, however, the veteran is connected with their old crew, that goes on another deployment, and lives are lost. "I should have been there," we tell ourselves. Certainly, our presence might have made a difference; and, just as likely, there was nothing that we could have done. Through no fault of our own, with no effort that we could have done that would have made a difference, things ended the way that they did.

And what has been done is enough.

The Myth of Invincibility

People forget that I'm a human being, just because I play a sport that everybody loves. We're human. We're not invincible. We share the same feelings and emotions that people on the outside feel. I don't think people really understand that — Terrell Owens

There is no such thing as an impenetrable tower or an invincible army. Time and toil will wear all things down. That's why the military is a young person's game; as we get older, we understand the lie of invincibility. Our heart tells us, our knees tell us, and sometimes our mind tells us.

In the military, we are told from the beginning that winning is everything. Invincibility is the goal, and the goal is achievable. Be strong. Show no weakness. The recruiting posters don't lie; everyone is a steely-eyed killer defending truth, justice, and the American way. I once read an article about the toughest man alive[22]...a retired Navy Chief Petty Officer who completed SEAL training, Ranger school, and Air Force Tactical Air Controller training. He's an ultra-marathoner, and once held the Guinness World Record for the most pull-ups in 24 hours. Feel inadequate yet?

The problem with believing in the myth of invincibility comes when we start to realize that we're not invincible ourselves, but think others are. That we think that because someone else has more than us, then they somehow have it all together, they have all the answers, they're the winners and I'm the losers. We fall into the trap of the comparison game. We start to let our belief in legends[23] get in our way.

We Hold Others Responsible for Not Meeting Our Unrealistic Expectations of Them

There is nothing that makes us angrier than realizing that out that our perceptions of someone do not meet their reality. General David Petreaus has an extra-marital affair. Pete Rose bet on baseball while a manager of the Cincinnati Reds. Tainted heroes, fallen from grace, simply because they did not meet our unrealistic expectations of them. Did they ask us to, though? Sometimes, maybe. Sometimes we present one thing to the world while knowing, inside, that we're something completely different. Often, though, we blame others for falling short of our expectations of them. "I'm disappointed in you" reveals more about our own beliefs about the person than their actions.

We Hold Ourselves Responsible for Not Meeting Our Unrealistic Expectations of Others

The other side of this particular coin is when we compare ourselves to others, and find ourselves lacking in some way. Keeping up with the Joneses, when we have no clue what's happening behind closed doors. We blame ourselves for not living up to some fantastical ideal, some unachievable invincibility that was built in our mind. When we go on Facebook and Instagram and see that everyone is meeting with the coolest people, at the best places, eating the most delicious and beautiful food. Life online is contrived[24]; we show each other our highlight reels while living our backstage lives.

Even what you're reading now isn't reality. If you and I were talking about this in person, these words wouldn't be the ones I use; having a conversation is very different from writing. It takes me an hour to put my thoughts into words in a coherent way, but I had four conversations yesterday about this very subject; comparing ourselves to the invincible myth of others. What that does to our thoughts, our moods. It can make us angry when others don't act the way we think they should, or depressed when we don't act the way we think others think we should.

Rejecting the Myth of Invincibility Reveals Humanity

The vast majority of the people we put on a pedestal didn't ask to be put there. They wouldn't want to be there if we told them. They would explain all of the reasons why they shouldn't be on that pedestal; just another regular guy or gal. And it's not selling ourselves short…it's revealing the accuracy of the situation. Superman bleeds, Batman loses his grip every once in a while, and the Lone Ranger couldn't get things done by himself. We are humans, and we do human things; like breathe, and stumble, and screw up, and feel bad about it.

If you feel yourself drawn into a negative opinion of yourself when you look at the lives of others, remember that we're all just trying to figure out this thing called life. We may put on a good show for others, but it's a game of escalating illusion; instead, revealing our flaws and acknowledging where we're not one hundred percent all the time is what makes us real.

I work with our local Veteran Court, and it has been selected as one of four courts in the nation to mentor other courts. Other groups from around the country come to our court to see how we do things. Our leader commits to showing everything we do…the good and the bad. He consistently reminds us to be completely transparent, "warts and all." If we're perpetuating the myth of infallibility, invincibility, then we're lying to ourselves and doing no good to others. Instead of delivering the beautiful lie, we should deliver the ugly truth, and let things be the way they are.

We could always do with more reality in the world.

Public Perception of Veterans and Veteran Mental Health

A colleague once told me a story of frustration that I've heard from many veterans in therapy. The colleague talked about a veteran he was working with; the veteran said, "People want to go to the movies and see the combat stuff…American Sniper and all that. But if I talk about the same thing at the neighborhood barbecue, they look at me like I'm crazy."

A lot of veterans I talk to feel the same way; they see the community rally around the concept of veterans, but don't always feel the same support for themselves individually.

We can say all we want that we support veterans returning from war, but do we really? Does the community rally around the flag and thank veterans for their service, but when it comes to real action, is it there? Unfortunately, there is little data to back up the prevailing wisdom of what veterans believe to be true. There do not seem to be studies on current public perceptions of veterans; there are some polls that show that perceptions have shifted from how people viewed Vietnam Veterans, of course, but actual research on the true public perception of military service seems lacking. Regardless of the research-based proof, however, I've heard repeated examples of the proof of this.

Thank You for Your Service

"Thank you for your service" represents the banality of society's understanding of the nation's wars and the men and women who fought them. – Marine Corps veteran Sebastian Bae, Heroes & Monsters: War's moral injury[25]

A Vietnam Veteran recently told me how it surprises him that current-era veterans don't want to hear "thank you for your service." In his mind, Vietnam Veterans would have *loved* to hear that when they returned home; but he also recognized that, while the words are there, they often seem meaningless. I serve as a mental health provider for our local Veteran Treatment Court. The presiding Judge, who is a combat veteran himself, has a statement that he tells every veteran entering the program. "Thank you for your service," he says. "I know that doesn't mean much from people who have never served, but coming from a fellow veteran, it means a lot."

My reaction is similar to many of my fellow Iraq and Afghanistan veterans. I do appreciate the intent behind the sentiment, but it bothers me. When I happen to bring up my military service in a conversation, it is sometimes to add context to what I'm saying or emphasize a point; but, in the middle of what I'm saying, I'm invariably interrupted with the "thank you for your service" line. Thank you, I appreciate it, but I don't want your appreciation, I want your understanding. Many veterans appear to feel the same way.

The NIMBY Effect

The concept of the NIMBY effect, which stands for "Not in My Backyard," is often applied to facilities, infrastructure, and services. We know we need a place to dump our garbage, but we don't want the town dump in the middle of the business district. Often, this makes sense; communities zone areas as industrial and residential for a reason. I grew up a mile away from the St. Louis Airport, and I can tell you, the sounds of airplanes overhead were constant. The NIMBY effect is described as opposition[26] to those things that are socially necessary, but have a negative connotation. Garbage dumps, substance abuse recovery centers…and veterans?

This goes back to the concept of "we love veterans…but from afar." The wider population knows, in general, that a standing military is socially necessary, and military service in general has a positive connotation…but veterans in the specific have a negative connotation. The stereotypes of Villain, Victim or Hero[27]. The crazy homeless combat vet.

Veterans in The Workplace

Veterans often hear of other veterans who struggle in the workplace due to common perceptions. I often describe it as, "We love veterans…unless we know one sitting in the cubicle next to us." Veteran Garrett Wilkerson describes his own experience with this in his article, "On Being 'One of Those Weird Veterans' in the Workplace[28]":

> …a coworker responsible for training me began using questionable language when explaining things to me. He would say things like "if you feel like you're going to have a *moment*, you can take a walk around the building to *cool off*". Do people here have "moments"? Is that a thing? On another occasion he told me that if I ever felt overwhelmed with all of the new information, I should be mindful to not "blow up" on other employees. I thought this was strange given that this coworker knew very little about me or my temperament, outside of the fact that I was a veteran.

Researchers have looked at this issue. In a study that looked at whether employers were prepared to hire, accommodate, and retain veterans with disabilities in the workforce, the research shows that wanting to support veterans is not enough. There are gaps between desire and action.

"Overall, our findings indicate that, though employers do have good will in this area, goodwill alone may not be enough to ensure that workplaces are geared up to enable [veterans with disabilities] to fully contribute their talents on the job." – Rudstam, Gower and Cook, Beyond yellow ribbons: Are employers prepared to hire, accommodate and retain returning veterans with disabilities?[29]

The fact is that the application of support for veterans does not always coincide with the expressed support for veterans.

The Veteran Role in Public Perception

I get the sense that this article may seem a bit rant-ish. And it probably is, because this is a frustration that I hear from many veterans in different areas. What role do veterans play in the public perception, however? The more the public misperceives veterans, the more veterans pull away. That simply contributes to the growing gap[30] between those who served and those who haven't. Just as with any other awareness, the individuals who make up the group that wants people to understand have a responsibility to conduct themselves in such a way that does not perpetuate the stereotype. In this way…stepping into the gap…we can change perceptions individually and as a whole.

The Benefits and Detriments of the Mission Focused Mindset

I was getting ready for the day recently, and I felt on edge. Ready. Anticipating. I had a lot of stuff to do that day, and I was ready to get cracking.

It was exactly the same feeling I had before the platoon went outside the wire as a security escort.

There is a certain type of live-wire energy, a physical vibration that runs through you when there's important stuff to do. I stood there for a minute, enjoying it. Reliving it. Then I realized…there's no life or death here today. It's phone calls and emails. There's no do-or-die mission, the world isn't going to end if something I have planned doesn't happen. So why am I feeling this way?

It's the mission focused mindset. My brain is telling my body to gear up to get this done. There are certainly some benefits to this type of mindset, but there are drawbacks too.

You Can Get Things Done with the Mission Focused Mindset

The one thing about being task focused: you get things done. In the military or out of the military, setting defined goals and achieving them was beneficial. Whether the mission is completing the obstacle course in record time, getting through tank gunnery, or getting a college degree, a goal-focused mindset is a huge benefit. Things had to get done when we were in the military, no questions asked. From getting to morning formation to start physical training to completing our part of an annual training exercise, you did the best you could as part of the team with the resources you had.

The mission oriented mindset was ESPECIALLY beneficial when it came to deployments and combat operations. The mission becomes the service member's entire reason for being. The satisfaction that comes with a job well done is one of the greatest feelings in the world. Combat operations are the ultimate test of individual and collective training, and lives are on the line. Being mission focused is what it's all about.

The Mission Focused Mindset Makes Us Trustworthy

Many of the veterans I know or work with pride themselves on being dependable. They like knowing that they can be counted on to get the job done. Again, the satisfaction that comes with knowing that, if someone gives me a job to do, then I'm going to do it. We can certainly take that too far, thought; that's part of the drawbacks of the mission focused mindset. Being dependable, however, is what service members are known for.

I had a First Sergeant at Fort Bragg that took a couple of the best Soldiers in my squad to perform other roles in the unit. I would then get some...challenging...Soldiers assigned to me. When I complained, respectfully, he told me, "You stop trainin' them, I'll stop taking them." While frustrating, it was also satisfying to be able to have the trust of your superiors. It's the same way in our post-military lives. If you want something done, give it to a veteran. It may not be the way you would have done it, but if the outcome is more important than the process that it took to get there, then a former service member will get you there.

The Mission Focused Mindset Can Lead to Black and White Thinking

One of the drawbacks to the mission focused mindset is that it can cause us to think in black and white terms. Pass/fail. Go/NoGo. Either you make it 100% or you don't make it at all. In a paper that looks at the types of metaphors[31] that veterans use in their post military life, Patrick S. Foley talks about the mission mindset. Mission orientation is equated to getting from point A to point B, no questions asked. Everything is reduced to whether or not the mission gets done; life is not as easy as that, of course. The type of all or none thinking can make us

I see it in myself. I have what I refer to as a "98%" mindset. There's an urge to do things at a 98% level (because nobody's perfect, right?). The challenge is when I get things done at a 90% level, it might as well be like I failed. Because it didn't meet the goal. I didn't achieve the standard...one that was probably stricter than it needed to be. On one had we're telling ourselves to take it easy, it's not that serious, but secretly we're pushing ourselves to be better, faster, stronger. Maybe it's even subconscious. But if we look at life as a pass/fail, black and white proposition, then we're going to be sorely disappointed. Life just doesn't work that way.

The Mission Focused Mindset Can Cause Us To Miss Important Things

This isn't just a "stop and smell the roses" thing, although it might be. In another article[32] looking at veterans in higher education, the mission oriented mindset was prevalent in most veterans. They wanted to get through school, in the least amount of time possible, within the parameters of their GI Bill, and get on to the next task. Again, I saw it myself. "What's the most amount of credits I can take in the shortest amount of time to check the requirements off the list?" Thinking that way, however, could cause us to miss out on some interesting experiences. It can also lead to burnout, where we lose focus on self-care, or our family and friends.

Serendipity plays a part in this as well. One of the key aspects[33] of taking advantage of opportunities that pop up through beneficial chance is having a mindset that is open to different ideas. We don't consider the side-paths when we're in a mission oriented mindset. We don't pick our head up, look around, and say, "what's that over there?" Because that would mean that we don't accomplish the mission in front of us…when we likely put there ourselves.

The mission oriented mindset is certainly beneficial, but like most things in our lives, can lead to negative consequences if not managed. Understanding how far to go with it, without going too far, can lead to balance in our lives.

Emotional Detachment in Combat and Coming Home

Call it survival, or becoming jaded or cynical. Many combat veterans experienced a point in their service when things that "should" bother them don't. It's hard to explain to someone who hasn't experienced it, of course. In "normal" life, we don't go up against situations that are horrific, horrible, or traumatic. Because they are so rare in our wider society, they are significant events when it does happen. In the military, emotional detachment is the name of the game, and it happens more than people might think.

It starts from the moment we enter the military. It's not about what we like and what we don't like...opinions don't matter much. It's about doing what you're told and not doing what you're not told. There are some benefits to this; for sure. It's often only when we're tested that we become aware of the capacity that we have[34] to endure stress and hardship. It continues when a service member gets to their first duty station and faces their peers and leaders. Certain military training...water training, Airborne school, obstacles courses...is designed to induce fear while increasing skill. This happens so that the service member can get the mission done while ignoring or decreasing the emotional response to these stressors.

And then there's combat.

When someone first enters a combat zone, they're not used to being in danger. Sure, they kind of know what to expect, but they don't really know until it happens. The sounds of combat may be disconcerting at first...but many combat service members get used to it. This emotional detachment...the apparent lack of fear, grief, or even sense of self-preservation...happens quickly. This type of detachment is seen in movies like Full Metal Jacket, as when Animal Mother[35] says, "You'd better flush out your head, new guy. This isn't about freedom; this is a slaughter" Colonel Kilgore in Apocalypse Now:

"Smell that? You smell that? Napalm, son. Nothing else in the world smells like that. I love the smell of napalm in the morning. You know, one time we had a hill bombed, for 12 hours. When it was all over, I walked up. We didn't find one of 'em...The smell, you know that gasoline smell, the whole hill. Smelled like...Victory."

It Was Useful in Combat

Here's a truth for many who haven't served: this type of emotional detachment was necessary. This topic came up for me after listening to a recent episode of a podcast series[36] about the WWII miniseries Band of

Brothers, hosted by my buddy and fellow podcast host, Jeff Adamec. In the third episode of the series, one of the unit's officers says this to a soldier who is having a difficult time adjusting to combat:

The only hope you have is to accept the fact that you're already dead. The sooner you accept that, the sooner you'll be able to function as a soldier is supposed to function: without mercy, without compassion, without remorse. All war depends upon it.

While this may sound cold, and morbid, and depressing, that's true; it's all of those things. And then the morbidity and coldness of that statement cease to matter. It's almost as if the statement could have been boiled down to two rules of combat: 1. Accept the fact that you're already dead. 2. When in doubt, refer to rule #1. It is when a combat veteran reaches a level of emotional detachment that they are able to accomplish the mission, whatever that is to them...covering their buddy's back, watching out for their own, all of it. Battlefield losses happen; someone is injured, someone is killed, and, of course, the war goes on, and life goes on. In many ways, it's as much a survival mechanism as anything else.

Although Useful, It Might Not Have Been Beneficial

Another painful reality of emotional detachment in combat: greater degrees of emotional detachment in combat might have an impact on the development of Posttraumatic Stress Disorder. In an analysis conducted in 2003[37], before the repeated exposure to combat that Post 9/11 veterans experienced, psychologists Matthew Tull and Lizabeth Roemer identified several studies that show that emotional numbing was the strongest predictor of PTSD five months after the traumatic incident. There is evidence that a service member that felt significant emotional detachment in combat would experience challenges regarding traumatic stress reaction upon returning from the combat environment.

Sometimes, It Doesn't Stay In Combat

The feeling of detachment often persists when the veteran returns home. I have found that this is especially true with veterans who were required to engage in multiple deployments. The emotional detachment of the battlefield is brought to the homefront, and when there is a likelihood that the return to the battlefield is less than a year away, the detachment is easier to maintain. When a service member is exposed to combat on their first deployment, it may take some time to get into "combat mode," weeks or even months. On the second deployment, it may take only a couple of weeks; in subsequent deployments, combat mode kicks in within days, or often even before the wheels of the plane hit the ground.

It Is Not Useful in Post-Military Life

Like the Violence of Action Paradox[38], what was beneficial to service members while they were deployed was less beneficial when the get home. The need to believe that you're already dead: while helpful in combat, not helpful at all when we return home to pick up our lives. When we need to reengage with others, establish or re-establish relationships, and start to care about things again, the habit of emotional detachment gets in the way.

So what to do about it? As with everything else, awareness and acknowledgement are the first steps towards making a change. We can't change something if we don't recognize it within us. After awareness, talking to someone to get understanding around it is extremely beneficial. Working with a mental health professional can be helpful, and it doesn't mean you're crazy; it means that you are using skills that kept you alive once, but are no longer necessary to accomplish the mission.

Just because you felt dead once, doesn't mean that you need to continue to feel that way. And coming back from that can make you appreciate life a whole heck of a lot more.

Preparation, Surprise, and Emotion Regulation

Forewarned, Forearmed; to be prepared is half the victory – Miguel de Cervantes

My family and I were having a great time. That awesome show, Mythbusters, was going into it's final season (with the original group). The hosts, Adam and Jamie, were doing a live tour. All of us are fans, especially my son and I. Mythbusters was one of the series that got me through my second tour in Afghanistan. The stage show was a lot of fun…until the moment it wasn't. During the big finale, they brought out an audience member dressed in a suit of medieval armor. He stood the edge of the stage; between him and the audience, they placed a transparent wall of plexiglass. Behind him, pointed directly towards the audience, they placed a howitzer made of paintball guns.

You can see where this is going for a combat vet.

When they let that thing off, the back of my neck tried to crawl out of the top of my head. I turned nearly sideways in my seat, gripped my wife's arm, clenched my teeth, and bore through it. The sound of paintballs hitting the plexiglass barriers sounded an awful lot like AK47 rounds hitting the windows of an MRAP. As long as it lasted…fifteen seconds that seemed like twenty minutes…I was straining to remain in my seat like a dog straining against a leash. Or, perhaps more appropriately, a dog reacting to a thunderstorm…and then it passed. As soon as it was over, I was able to calm myself down, shake it off, and go on with the night.

Preparation Versus Surprise

There is something about the unexpected that always throws us off our feet, especially when we're surprised by something negative. I've had a lot of discussions lately about preparation and surprise. I recently had the privilege of attending a talk by Kevin Sullivan[39], George W. Bush's communications director for two years; when talking about interacting with others, he said, "preparation is good, surprises are bad." Similarly, in a recent podcast episode[40], Dr. Carmen McLean from the national center for PTSD talked about the difference between being prepared for something negative versus and being surprised by something negative. She said that a colleague always told here, "There is a difference between someone telling you to get ready to catch a ball, and someone walking through the door and beaning you with a fastball."

The Neuroscience of Surprise

Science bears out the idea that being surprised by a negative stimulus is a bad thing. In a study to determine the neurological reactions to adverse events[41], researchers found that predictability controls the response of the

50

amygdala to both pleasant and adverse events. If we are surprised by something positive, the amygdala amplifies the positive feeling; if we're surprised by something negative, the amygdala amplifies that, as well. The amygdala, small, almond-shaped structures in our brain that control the positive and negative emotional reactions to events, has been proven to be a critical component to the way our brain reacts to traumatic events[42].

For many veterans, both with and without PTSD, increased activation in the amygdala is a result of repeated exposure to traumatic events. The amygdala is always "switched on," accounting for the increased hypervigilance that many veterans experience. Couple a surprising negative event with a portion of your brain that is already operating at a high level, and you get the kind of response that I had to a paintball howitzer.

Negative Events Are Easier If You're Prepared

In contrast to the amygdala controlling your emotions, the prefrontal cortex..the front part of our brain...is involved in emotional control. Consider the amygdala and hippocampus as the "lower" parts of our brains, and the prefrontal cortex as the higher front part of our brain. I often explain it to the veterans I work with in this way: the lower part of our brain is the gas pedal, which revs the engine when the green light goes on. The front part of our brain acts more like a governor in an engine, a speed limiting device that keeps the engine at a safe operating level.

If we are prepared for a something negative, then our prefrontal cortex is already engaged. That's what Dr. McLean is referring to when she says, "I'm going to throw you a ball. Okay? Get ready…" When that happens, our brain gets prepared for the event, and the impact of the runaway amygdala is reduced. We are able to react more coherently to an event if we have an understanding that it's about to happen. Consider the paintball howitzer event above again; recalling the story as I write this, I have none of the reactions that I did when it happened. Because I was prepared.

Being Prepared Doesn't Always Mean Being On Guard

There's a difference between prepared and always being on guard. We can't walk through life always thinking the worst is going to happen, because that's what we're always going to see. It's as if we are always walking around with an umbrella, waiting for the storm. If it rains after three weeks, we tell ourselves, "Ha! I was ready for it!" What we don't realize, though, is that we missed three weeks of beautiful weather waiting for the rain.

Sure, sometimes someone is going to walk in the room and throw us a curveball. We may not be ready for it, and if we've experienced a bunch of negative events in our life, our brains will react in a big way. But if we can't engage the front part of our brain before the event, we can certainly do so quickly afterwards. The paintball howitzer? I didn't let it ruin my night. It was a sudden storm, violent and quick, but it passed just as rapidly as it came. I didn't throw the family in the car, rush home, and lock us all in a bunker. I shook it off, took the kids out for some ice cream, and chalked it up to the life of a combat vet.

Humans have the amazing ability to control our own emotions; we just have to get out of our own way to do so.

Weighed Down by Past and Future

With much of the work that I do with veterans, it's like we're still carrying the same loads that we did when we were in the military. The heavy rucksack, the duffle bag drag, the weight on our shoulders…and our mind, our souls. The burden of these weights can be so great that we come to a standstill and get stuck. They can be so great as to be defeating, crushing; the weight that we carry can be so great that it sometimes kills us.

What is the weight that we carry? It certainly is some of the things I've talked about in the past[43], such as the names and places we've been to, or the people we served with. The people we lost. But often, it's the weight of things from our past, and weight from a future that may never happen. We burden ourselves with things that are not our responsibility, don't belong to us, and that we can do nothing about. Talk about an exercise in futility…

Future Weight

My life has been full of terrible misfortunes…most of which never happened – Michel de Montaigne

"I know how this turns out; no matter how hard I try, it's all going to end up in the crapper." I can't count how many times I've heard variations on that theme. Or, "why bother? It's all going to blow up in my face anyway." Sometimes, we pull the future towards us and bring it into the present. We create doomsday scenarios that defeat us even before we get onto the playing field. The weight of this future is so heavy, the prospects for success are so slim, that we shouldn't even try.

If we're so good at predicting the future, why is it that we often see only catastrophe? The problem with reacting in the present to a potential horrible future is that it will increase our awareness and sensitivity to pain. In a study conducted in 2001[44], individuals who scored high on a measure for catastrophic thinking anticipated that an ordeal (putting your hand in a bucket of ice water for one minute) would be more painful than those who did not score high in catastrophic thinking. The results showed, however, than the catastrophic thinkers rated their pain *even higher than they anticipated.* In other words, they thought it would hurt, and it hurt even worse than they thought. Self-fulfilling prophecy? Absolutely. Being weighed down by doomsday scenarios in the future hurts us more in the present AND in the future. Why do it?

Past Weight

"Sooner or later she had to give up the hope for a better past." — Irvin D. Yalom

We've all heard it so often before, that the phrases bounce around in our mind like empty balloons. "No man steps into the same river twice." "We will never walk the same path again." The past is the past, and that is incontrovertible truth. Nothing we can do can change what happened twelve years ago, or fifty years ago. And, yet, we are so bogged down by the weight of past failures, past regrets, past mistakes, that we are essentially immobile in the present. We base our opinions on past experience; so if we carry the weight of the negative past, then the prospect of the future will be just as negative.

The past will not be any better or worse than it was. It is simply things that happened in our past. If we put negative meaning to it, however, it's going to bog us down and make us not even want to get out of bed. It's going to make us not even try for future success, because past success has been so elusive. Constantly thinking and worrying about the past…rumination…is proven to be associated with negatively biased thinking, poor problem solving skills, and increased negativity, as shown another 2001 study [45]. So if we're weighed down by the past, and are poking at it like a bruise that never fades, then we're more likely to look at the future catastrophically as well.

Offload the Weight

When we were in the military, and we got done with the ruck march, we didn't walk around with the weight for the rest of the day. Did we say, "No thanks, Corporal, I think I'll just keep carrying it"? Heck no! We dropped that crap as soon as we possibly could. And felt relief after doing it, too. So why do we do it with our mental and emotional weight? Why do we still stuff our pockets with failure? Why do we load regrets onto our back, put on catastrophe glasses to look at the future? Is it fun, enjoyable? Of course not. The thing is, we can change it. We can change our outlook through deliberate effort. We first have to realize that this is what we're doing. Then we have to want to change, and believe we have the ability to change.

Only then can we offload the weight and get on with our day.

When the Cure Becomes the Curse

When a service member returns from combat, or leaves the military, they take their experiences with them. In an episode of the Head Space and Timing podcast, Dr. Larry Decker[46] said, "The lessons learned in combat are never forgotten." It's true; I know this both as a clinician and as someone who learned a thing or two while deployed to Iraq and Afghanistan. There are some things that we're not going to forget, if we're blessed enough to avoid dementia or Alzheimer's disease.

There are things that many veterans do want to forget, though. I can't count how many times I've had a veteran say to me, "I wish I could just flip a switch and get rid of this stuff. Or delete some files." Of course, both veterans and mental health professionals know that it's not that easy. Memories, especially traumatic memories, are encoded in our brains. The military training we internalize...push through the pain, advance at all costs, remove the emotion from the equation and accomplish the mission...get us in the habit of suppressing that which is uncomfortable. Unhelpful. Detrimental to achieving the objective.

So, in the absence of a switch or a system-wide reboot, we turn to other things to do that for us. Alcohol. Mind and mood altering medications. Relationships, some of which may not be helpful to us. Even behaviors that help us get through the day. The dangerous thing about these coping techniques: they really work. In the short run. The thing that gets us through the day-to-day problems, the cure to what ails us, becomes a curse in the long run.

Substance Use and Addiction

Addiction is a significant problem for service members and veterans. In an article titled Substance Use Disorders in Military and Veterans: Prevalence and Treatment Challenges[47], the authors identified that over 80% of Post-9/11 veterans diagnosed with a substance use disorder were also diagnosed with another mental health condition. This could include PTSD, Major Depressive Disorder, Generalized Anxiety Disorder, and others. One possible conclusion? In order to cope with the discomfort of traumatic stress reaction, depression, or anxiety, service members and veterans turn to substances. Alcohol turns things off for a while. Painkillers make us not feel, one pill to pick us up, another to calm us down.

Like I said before, they really work. They're effective, or else we wouldn't use them. I've spoken to many veterans who saw their drinking behavior change; before they deployed to combat, sure they drank. It was a social thing. Celebrating with the buddies. For many young service members, this is the first time that they're away from home. Many college students experience the same thing. After deployments, however, many veterans found that they were drinking not to make themselves feel better, but to keep them from feeling worse.

Dysfunctional Relationships

Many veterans have heard this: "You've changed since you came back." Of course; everyone changes as a result of their experiences, not just the veterans. The kids, the spouse, the parents and friends; everyone experienced a disruption in their lives that takes a toll. One of the things that isn't taught in school is how to have a healthy relationship; we learn from watching others do it. We learn how to have romantic relationships from watching our parents, our friends, and (of course) portrayal of relationships in the media. On top of this, put how we manage relationships that have changed due to the pressures of military service.

Distancing ourselves from relationships is another behavior that works well in the short term; it does what we need it to do. It keeps us from having to answer questions. Avoiding others shields us from the disappointment and hurt that we see in someone else's eyes, which sometimes reflects the disappointment we feel in ourselves. In another article, Military Related PTSD and Intimate Relationships[48], the authors identify that intimate relationship problems that go along with traumatic stress reaction can influence the course of recovery. Pushing loved ones away, isolating, terminating the relationship altogether; each of these works in the short term to resolve immediate discomfort, but cause significant challenges in the long term.

Maladaptive Thoughts and Behaviors

Maladaptive thoughts and behaviors are those that we engage in that keep us from being able to adjust to a particular situation. We engage in them because they work…in the short term…but they ultimately keep us from resolving the main issue we're trying to deal with. This is contrasted with adaptive thoughts and behaviors, those that help us resolve the issues we're facing and adjust to them. There are a number of maladaptive behaviors[49] that service members engage in: avoidance, self-harm, eating disorders. Engaging in substance use, as described above, is a maladaptive behavior. Again, these coping techniques tend to work, or we wouldn't use them; however, they have long-term and dangerous consequences.

Thinking errors and cognitive distortions are equally damaging to us. In an article by Amy Morin[50], author of 13 Things Mentally Strong People Don't Do, a list of common thinking errors include catastrophizing, over generalizing, mind reading, and others. Again, we engage in them because they meet a need. They calm us down (in a sense), they help us make sense of the world around us. The challenge, as always, is finding a balance between what we think and do that makes us feel better and what is most effective for us in the long term.

Seek a Long Term Cure

When we're looking to address the challenges that come along with our military experiences, veterans can and should seek out those solutions that work well in both the short and long term. What is effective both now and in the future. That's when we move from disruptive adaptation to our experiences to successful adaptation to our experiences. That is what leads to the post-military life we desire and deserve.

Insomnia and Veteran Mental Health

Service members and veterans have a unique relationship with sleep. On one hand, troops can fall asleep at the drop of a hat. Standing up, leaning against each other, airplane, ship or tank...we could catch Z's any time we wanted. The old joke in the motor pool was that you could snatch a few if you crawl up under a vehicle and hook your hands into the undercarriage...it looks like you're working, while you're not.

On the other hand, we didn't do enough of it, and what we did get was not very restful. You're not getting a lot of stage three deep sleep bouncing along on the inside of an aircraft, waiting to jump out of it. We often prided ourselves on being able to function on minimal amounts of sleep, field by caffeine and edgy irritation.

The complicated relationship that we have to sleep carries over into our post-military lives.

One of the most universal challenges that I see in veterans I work with are sleep problems. As in, they don't. Insomnia is a common condition that many veterans experience, for a variety of reasons, and it can wreak havoc on our psychological wellbeing as well as our physical health. Insomnia and mood disorders can go hand in hand, and particular symptoms of PTSD are sleep-disrupting as well. There are treatments and supports, however, and if we can figure out the sleep issues we have, we're ahead of the game when it comes to managing our moods.

Anxiety and Depression

For starters, many don't realize that insomnia doesn't just mean "I can't sleep." There are different forms of it. In this research study[51] looking at the intersection of insomnia, depression, or anxiety, the three kinds of insomnia are Onset (can't get to sleep), Maintenance (can't stay asleep) and Terminal (wake up early and can't get back to sleep). Anxiety disorders are associated with onset insomnia; "I can't get to sleep because I'm worrying about tomorrow" or "my mind just won't shut up." Depression is more related to terminal insomnia; you can get to sleep pretty well, but wake up at two or three or four in the morning. Maybe waking up at four was the norm when you were in the military, but when you've been out for ten years? Different habits should have taken over.

The problem is that there is a chicken-and-the-egg thing going on with insomnia and mood disorders. Studies have shown[52] that service members with greater levels of insomnia before they deployed are more likely to develop depression and anxiety, and that insomnia four months after deployment was a predictor of depression twelve months after deployment.

And it doesn't just go away; a study in 2015[53] found that depression, PTSD, and insomnia were closely linked in Vietnam veterans…forty years after they left the military. It also becomes a cycle. You can't sleep because you're anxious or depressed, which leads to an inability to manage your anxiety and depression, so you're more anxious and depressed. So you can't sleep. The mood disorders become associated[54] with the bed and the bedroom, so that the two are as linked as standing at attention when Reveille sounds off.

Hypervigilance and Nightmares

On top of the impact of depression and anxiety on sleep, veterans also experience sleep disturbances specifically related to traumatic stress reaction. Hyperarousal, increased responsiveness to things that happen in our environment, is a primary symptom of Posttraumatic Stress Disorder. The perpetually keyed-up state that our bodies are in as a result of hyperarousal make it hard to rest and down-regulate our nervous system…which makes it hard to sleep. This is more of the chicken-and-the-egg…is the insomnia make the hyperarousal worse, or does the hyperarousal make the insomnia worse? Studies indicate[55] that both could be true. I've had veterans say that every little noise wakes them up, and they're up and around the house, ready to go. Sound like the troop sleeping on the ship, vehicle or aircraft? Old habits are hard to break.

And then there's the nightmares. You ain't getting to sleep if there's a horror movie on every time you close your eyes. Specifically, the nightmares that disrupt the sleep of many veterans is the vivid reexperiencing of military-related traumatic events. This is a constant re-exposure to the distressing events, which perpetuates the traumatic stress reaction symptoms, but it also disrupts sleep to a significant degree. This review of evidence and studies[56] on insomnia and nightmares in 2012 shows that insomnia and nightmares for veterans experiencing PTSD are linked.

Treatment and Support

I've found that insomnia is one of the things that can really get a veteran to engage in treatment for their psychological distress. There are less beneficial ways to address this…I've heard a veteran say a twelve-pack a night is what was able to get them to sleep. Effective, sure but helpful? Absolutely not. Eventually, the cure becomes a curse. There are a number of different types of treatment specifically designed to address insomnia.

Cognitive Behavioral Therapy for Insomnia (CBT-I)[57] is an intervention that has been used with success, is recognized by the VA, and is considered a primary intervention for insomnia by the National Institute of Health. It has been shown to improve symptoms of anxiety and depression, and shows promising results in reducing PTSD symptoms. Imagery Rehearsal Therapy (IRT)[58] is another form of cognitive therapy that has been shown to be effective in reducing nightmares specifically. This is a specific set of treatments that first help the veteran to recognize that the nightmares promote learned insomnia, then to learn how to control our mind's imagery system. IRT is essentially learning how to change a nightmare into a new dream while you're awake, so that your mind does it automatically when you're asleep.

Of course, there are the normal sleep hygiene things that go along with getting good sleep. Use the bed for sleep and sex. I get it, many will read before sleeping or watch TV in bed, but cognitive stimulation is the opposite of what we need when we're trying to get to sleep. Don't drink coffee at 8pm and expect to get to sleep later. You can see a number of different aspects of sleep hygiene in this article[59]. And, again of course, there are medications that help reduce nightmares and induce sleep. Not everything works for everyone, but talking to your medication prescriber about the type and frequency of sleep disturbances you have can be beneficial as well.

Because we all know what it's like to work on no sleep. It wasn't fun when we were in the military, and it's certainly not fun now; figuring out this part of post-military life can greatly improve the quality of everything else.

Does Time Heal the Psychological Impact of War?

No matter what happens in life, whether we think it's good or bad, try not to evaluate it in the moment. The truth is, we don't know whether it's truly good or bad until we get some distance from it. - Kevin Maney

It's a statement that I've heard many veterans say: "I was told to just give it time, things would get better." Is that really true? There is something very powerful about the passage of time, of course. Hindsight is 20/20, distance gives us perspective, and all that. But just allowing time to pass, does that make things that happened any easier to bear?

As I consider the quote above from the lens of veteran mental health, I can see that it applies. When we get distance from an event, we see how that event impacted our life. It's very hard to evaluate the impact of an event as you are experiencing it, or even removed from it by a couple of months or years.

Don't get me wrong; the loss of our brothers or sisters, or a significant traumatic event, is never going to magically turn into rainbows and puppy dogs, no matter how much time passes. For example, the picture at the top of this post; no amount of time will lessen the impact of 9/11. And I will never say that "it's a good thing that this happened" to traumatic loss. What time does, though, is give us perspective about these events, and an opportunity to understand them.

This can apply to other really horrible situations, though. As I've mentioned before, I work with justice involved veterans. One of the common statements I hear is, "it really sucks that it happened, but it was probably the best thing for me." Here are some other thoughts on how the passage of time can provide both perspective and relief.

The Passage of Time Reduces Emotions

It is very, very difficult to maintain a heightened state of emotional response for an extended period of time. The human body is not designed to operate for very long on extreme rage, severe terror, or debilitating depression. Emotions are critical for our interaction with each other, for communication, and even for survival. We wouldn't have them if they weren't necessary. Like many necessary things, however, too much of a useful thing could be counterproductive. If we react to a perceived great thing with positive emotions, be careful: it might not turn out the way we think. How many times have we expected something good and it turned out bad? Of course, we don't have to always assume that the worst will happen either, but time passing will allow our emotions to subside.

When I was in the military, one of the things I tried to always do is not to make decisions during moments of extreme emotion. When I found myself starting to lose my temper (or, more often, someone brought it to my attention that it was already lost), I tried to take a step back and not make any hasty decisions through the fog of emotion. It didn't always work...you can ask some of my joes...but, by and large, when I was angry, or frustrated, or down, I waited to make decisions. Those decisions I made in the heat of emotion were more likely to be poor ones.

The Passage of Time Allows Us To Apply Logic to Emotion

When considering an event, allowing time to pass lets the emotion cool down and the rational, reasonable side of our nature to take over. I've talked before[60] about Dialectical Behavior Therapy; one of the first concepts that I teach veterans about DBT is how to identify states of mind[61]. We have an Emotion Mind, which is based on our mood and focused on our emotion, and we have a Reasonable Mind, which is rational and task-oriented. When a significant event happens in our life, good or bad, we are likely to respond with emotion mind. When time passes after that event, we often use reasonable mind. The key, however, is to react to events, and life in general, with wise mind: the balance between the two.

I often describe Emotion Mind and Reasonable Mind in this way: consider Sherlock Holmes, cold reason and cool logic, compared to Dr. Watson. Dr. Watson is passionate, impulsive, quick to anger and jump to conclusion. Holmes is Reasonable Mind, Watson is Emotional Mind. Or, consider it this way: Captain Kirk and Mr. Spock. Kirk is all impulsive emotion, Spock is rational restraint. In both of these pairs you have opposites that complement each other; if left on his own, Holmes might figure it out, but miss the humanity in the problem. Kirk, if left to his own devices, would let his gut (or other things) get him into trouble that logic could have kept from happening. It is the combination of the two of them together, emotion and reason, that makes the teams truly effective. The passage of time allows reason to counterbalance the emotion.

The Passage of Time Gives Us More Information to Consider

One thing that we get with time is the ability to reflect on the event, and consider how subsequent events have played out. When it comes to making sense of trauma, though, time is not just enough to heal. I have seen veterans who have entire decades pass, and they are just as frustrated, angry, and bitter about things that have happened thirty years ago as they are about something that happened yesterday. Do we really take the time to consider things in the intervening time, or are we still caught up in the emotion of the original event?

Simply allowing the locked footlocker of our mind to remain, unopened, with all the horrible crap in there that we don't want to think about, isn't going to heal things. Opening it, unpacking it, keeping what's important and throwing out what's not helpful...that's the benefit that time gives us. That's why we shouldn't react in the moment...because it's the moments after that truly define the event.

The Minefield of Painful Memories

We all have painful memories in our lives. Loss. Regret. Mistakes and failures, real or imagined. It's part of the human condition, and I don't know anyone who's immune to it. Those who served in the military, though, have had more opportunity to accumulate painful memories. I'm not talking about when the Drill Instructor yelled at you, or when that gal or dude you were seeing broke up with you. That's small potatoes when it comes to the really big stuff. Combat, mistakes that were made by me or someone else that cost someone their life. Things that a service member or veteran saw that can't be unseen. Navigating through life with these memories inside our heads can be like walking through a minefield. Dangerous, but possible, if you know what to do.

The meaning of the minefield of painful memories in post military life can be significant. Our emotions, our relationships, our careers…all can be impacted if we don't address these painful memories and put them in the proper place in our minds. This doesn't mean burying them and ignoring them…like real minefields, ignoring them and pretending like they're not there simply doesn't work. Instead, taking the time to understand, to process, to reduce the impact of the painful memories is the surest way to move beyond the minefield.

You Can't Get Through The Minefield Without Knowing You're In It

Bosnia, 1996. My unit was transporting supplies from one base to another when we ended up on the wrong route. No Lieutenant jokes, please, although there was one leading the convoy. We started seeing signs in Cyrillic, meaning we were heading into Serbian territory…no bueno. We start to turn around in a wide area near the road, when this old man came running out of a nearby house, waving his arms…the lead truck had pulled into a minefield. Slowly, carefully, they pulled back out in the same tracks they made pulling in. We didn't have awareness that we were in the minefield…but the old man sure did.

The problem of being stuck in a minefield is that you don't know that you're in it until something blows up. It's the same thing with painful memories. We think we have it figured out, it was all in the past, and it doesn't matter anymore, anyway. Except…it does. A lot. Even if it's only certain times of the year that it impacts us, we're still changed by the experience. So the first step in navigating the minefield is to become aware of the fact that we're in the middle of it. To start recognizing the signs before we get hit with a painful memory, rather than recognizing the aftermath.

You Don't Navigate The Minefield by Avoiding It

One of the most common ways of dealing with a minefield is to avoid it all together. Works great, if you can do it. In the story above, we didn't need to turn around in the field; we simply backed up until we found a parking lot big enough, and used that. The minefield of painful memories, however, can't be avoided. It stretches across our path, blocking the way to the clear land on the other side. There's no way around it; the only way to get beyond the minefield is through it. Again, we can blunder our way forward, doggedly absorbing the blasts until we finally make it to the other side; or we can find a way to navigate it safely.

If we continuously reroute ourselves around off limits areas that we place in our lives, we will never truly get to a place of peace. If we start down this path, then the painful memory pops up…and we avoid it. By ignoring it, denying it, burying it in distractions or drowning it in substances. We back off and try to find another route. And we keep coming up against it. The way beyond the minefield is not avoidance…it's acknowledgement.

Navigating the Minefield is Possible

"Why would I bring that up? I want to forget it, not relive it." In the meantime, we're experiencing problems in our post-military life. Substance abuse. Uncontrollable emotions. Disrupted relationships. Pretty much all of the factors of the comprehensive veteran mental health model[62]. If we don't navigate the minefield, then we don't resolve these challenges, either. The key, however, is that navigating through, and beyond, the minefield is possible. Maybe it's possible on our own, we make it through by luck or by sheer determination. It may be easier, however, to have someone help us get through it. By showing us where the painful memories are, by carefully defusing them and removing the danger. By understanding how the painful memory got there, what it's doing to us, and why we react the way we do.

It's not that we're broken, or crazy, or a monster. Service members have seen and done stuff that's outside of the realm of experience of our neighbors, if they haven't served. We're not perpetuating the stereotype of the broken warrior, any more than addressing the psychological impact of an automobile accident or natural disaster makes someone crazy. It's simply the truth of our experience. Painful memories can impact us long after the events have happened, and if we don't acknowledge and address it, then we can't move on.

And being stuck on the painful side of a minefield is a dangerous place to be.

.

Avoiding Pain Because of the Fear of Suffering

Out of suffering have emerged the strongest souls; the most massive characters are seared with scars — Khalil Gibran

I hear it often from the veterans I work with. "I don't want to deal with it, because if I do, the pain will never stop." Avoidance...of a topic, of a memory, of a story, of something that reminds us of trauma...is common when it comes to veteran mental health. It's a key diagnostic aspect of PTSD. We avoid feeling pain because there is a fear that we will continue to suffer.

But is there a difference between pain and suffering? Does one inevitably lead to another?

Physical and Emotional Pain

Pain can be both emotional and physical. We all know that physical pain, especially service members and veterans. The pain of pushing your body beyond it's known limits. Pain that comes from a twelve mile ruck march. The pain in the knees, back, shoulders, feet, neck, everywhere. The marginal "suck it up and drive on" pain and the excruciating "I'm not hurt, I'm injured" pain.

We're also familiar with emotional pain. The stinging rebuke from a respected leader. The sharp agony of failure, the crushing blow of defeat. There's also the deeper, longer pain; separation from family, an ended relationship, and certainly the loss of a brother or sister. These events cause pain that is as real as the physical.

Difference between Pain and Suffering

Suffering is pain, but it's never ending. Enduring, long-lasting. Pain can come and go, but if it remains for an extended period of time, then it turns into suffering. Again, this can certainly be physical pain; a lingering injury, or the chronic pain that many veterans experience. Chronic pain is medically defined as any pain lasting longer than twelve weeks; three months. Again, many veterans have been there, "sucking it up and driving on" for longer than that, which often leads to more long-lasting damage.

Emotional suffering is similarly long-lasting. The pain of loss is common; to still feel the pain of that loss, at the same intensity and with the same reactions, three or six months later; that's the point that it has turned into suffering. Often, we're not suffering deliberately; we don't even know we're keeping the emotional pain alive with our perceptions of it.

Suffering Is Not Inevitable

When it comes to emotional suffering, we often have the ability to manage it. This isn't the "suck it up and drive on" method of dealing with pain, but instead recognizing that our pain has turned into suffering. Acknowledging it. Doing something about it; talking it over with a trusted friend or mentor, or, if the suffering is significant enough, talk it over with a mental health professional. Hopefully we would do that when it comes to physical pain; again, not a typical response to the push-forward-at-all-costs mentality of our military selves, but if we felt physical pain at the level of some of our emotional pain, we would have been on the doctor's doorstep a long time ago.

Avoiding Pain out of Fear of Suffering

So if pain is pain, and suffering is unending pain, why avoid it? No-brainer question, of course. Who would willingly submit themselves to pain, if we knew it led to suffering? That's what we do, though. There is a painful memory or event in our past; an incident that changed us. Trauma. The memory of an event that still haunts us. Why would we even consider thinking about it…that would increase the pain, right? That would intensify it, prolong it, keep it going…cause us to suffer. Right?

The opposite is true, actually. I've seen it over and over again. The veteran thinks, "if I open that box, I'll never get it shut. Do you think the nightmares are bad now? Wait until I start talking about it." However, when we do start talking about it, understanding what happened, describing it to ourselves, then the pain actually goes away.

Relieving Suffering Through Pain

The thing is, even if we avoid dealing with the source of emotional suffering, we are often suffering. Avoiding the pain of remembering, thinking about it, talking about it, actually causes us to continue to suffer, not keeps us from suffering. It's only through the pain…not avoiding the pain…that true relief from suffering occurs.

Suffering is about what we tell ourselves. "If I start talking about it, I'll start crying, and I'll never stop." Really? Never? Forget the fact that veterans don't cry. Male or female, it's not something veterans often willingly allow ourselves to do. But to never stop? Physiologically impossible. If we tell ourselves these things, and believe them, then they're going to be true…and we will be continuing our suffering.

Relieving Suffering Does Not Mean an End to Pain

So here's the kicker. Just because we stop suffering doesn't mean that we won't feel pain. This is another fear that many veterans have; if I come to terms with the loss of my brother or sister, and stop suffering for it, then it means I don't care about them. I HAVE to continue to suffer, or their deaths would be meaningless. Consider it from the other side; you obviously care about the person you lost. If they were alive and you were gone, would you want them to suffer endlessly? To be in excruciating emotional pain? Of course not. You would want them to honor your life with their best life well lived; and that's what your buddy would have wanted for you.

No, dealing with the painful memories will not remove the pain entirely, but it will likely alleviate the suffering. The pain will not be long-lasting and constant. It will not be a daily burden, impacting all areas of your life. Instead, it will be a periodic pain, like your knee that twinges when a storm is coming. It will be manageable; it will remind us, occasionally, of the loss of an honored friend, but it won't impede every moment of our lives.

And that's a pain we can live with.

Sharing Painful Stories

Many veterans have been carrying the load of their story for years. Their secret. A heavy weight, burdening their mind, their soul. It doesn't mean veterans are crazy. It doesn't mean they're monsters; far from it. This weight means they're human, with a human reaction to inhuman events.

What veterans saw, especially in combat, opened their eyes to a world that they wish didn't exist. What veterans experienced exposed the worst demons of human nature, and sometimes exposed the worst of our own. Why would someone want to share that, to expose that to others? It should be forgotten, buried, left alone in the wilderness to wither and die.

Except that our brains don't work that way. With traumatic memories especially, the memory doesn't just fade into the background, like the name of that dude in high school you can't recall. They tend to linger, growing in shape and mass, and come out in other ways. And are dealt with in other ways; avoiding, in the form of denying they happened. Suppressing them, drowning them in work or alcohol or anything else that will keep us from thinking about them.

And yet the stories remain.

Then there comes a point where enough is enough, keeping it inside isn't working; it's still bouncing around inside of our head like a rubber bullet. These aren't dinner party stories, though; we can't just come out and tell anyone. They're not the things we tell around the Thanksgiving table or the neighborhood barbecue. They're the stories that, if we heard others tell them, we might react with shock or surprise.

But we don't. Because we've been there.

Telling Your Story Isn't Easy

This isn't the "the time I was in the military" story, or the story of a veteran's deployment...what it meant to them, how it changed them. No, this was a specific story, a moment in time, an event that occurred. These are the stories buried within the stories, like nesting dolls, deeper in and more closely held. We don't want to tell those stories to others; heck, we don't even want to tell those stories to ourselves. We often don't realize that we are avoiding it at all; in a form of Orwellian double-think, we try to forget the memory, then we forget trying to forget. It's not easy telling the stories in an emotional sense, in that it will make us feel bad; it's also not easy in the sense that even the effort of recalling the memory takes work. It takes psychological exertion to deliberately choose to move towards telling the story while everything within us is dragging us away from it. Moving towards telling the story is like moving towards the sound of gunfire.

Telling Your Story takes Courage

For that reason, telling the story of one of the worst days of your life takes courage. It's not weakness to admit that there was a time when you saw something, or did something, that still impacts you today. It's strength. The military mindset is one of minimizing weaknesses, eliminating threats, getting rid of anything that may get in the way of accomplishing the mission. In many ways, that's avoiding fear. Avoiding pain. Sucking it up and driving on. Which is extremely effective…until it isn't. When confronted with any physical obstacle in our service, we moved toward it, not away from it. We can use that same determination to move toward the painful story, rather than avoid it.

It also takes courage to trust someone with the story. As I said, this isn't dinner table conversation. This isn't talking about sports or the weather or that person you hooked up with last week. These aren't precious secrets that we want to keep just to ourselves; these are the stories we want to lock away forever. To share that with someone, to have a space to share that with someone, is often difficult. So difficult, in fact, that we do everything in our power to avoid it.

Telling Your Story Eases the Burden

Just like with any effort, however, the reward is at the end. The feeling you get after a hard workout, or a long effort on an important project. The feeling of relief that comes from finally setting down the rucksack or seeing the finished product. The sense that the effort that it took to get through it was worth it, that there is a benefit to completing it. The same thing happens when we choose to open up, to tell our story in a safe and trusting way. It works. I've seen it work; it's worked for me. When we avoid avoiding, we can approach understanding in a way that isn't possible if we lock the story inside.

The Post-9/11 Veteran Divide: Success or Struggle

The veteran community is at a crossroads, and there is a growing divide. This is not the civilian-military divide[63], although that certainly still exists, but a divide within the veteran community. This divide is between those veterans who are succeeding in their post-military lives, and those who are struggling. Between those veterans who are making a positive impact on their lives, many times in spite of their experiences in the military, and those whose service is having a negative impact on their lives.

The crossroads is the group of veterans who are becoming this century's Greatest Generation, and those veterans who are becoming this century's Lost Generation.

Many are familiar with the post-World War II generation, the "Greatest Generation." I've written before[64] on the possibility, and even the responsibility, for veterans of today's era to mirror the impact on our nation by that generation. In the fall of 2016, there was an excellent article in Task & Purpose[65] outlining the potential of this generation to become the Lost Generation; a generation of combat veterans who return to a war-weary country that is simply wanting to move on from the impact of a global conflict.

The Growing Divide Reflects the Generational Diversity of the Post-9/11 Veteran

At the time this is written, conflict in Afghanistan is moving into it's 17th year. Again, I've written that this is the first cross-generational conflict[66], but consider the senior leaders at the beginning of the war; many of them had been in the military for 25 years-plus. They were Baby Boomers, the generation that remembers the Korean War (generally) and definitely remember the Vietnam War, and possibly even served in military during that time.

The youngest Post-9/11 combat veteran might not have even joined the military yet, and the post-Millennial generation has been old enough to join the military for five years. So the Post-9/11 veteran could potentially be reflected in four generations: the Boomers, Generation X, the Millennial Generation, and Generation Z (or whatever name emerges for the post-Millennial generation). Put another way: the most senior leaders of the Post-9/11 military at the beginning of the conflict were likely born in 1949-1951. The youngest service members deploying today were born in 1999-2001. That is a span of a half a century.

The Growing Divide Is Impacted by Technology

The numbers of veterans, much less combat veterans, are nothing compared to the numbers of service members in either WWI or WWII. What is different in this generation, however, are advances in the medical field, digital technology, in warfare as a whole. More is being done with less people because the technology and the capability of today's forces were unimagined a century ago.

In the same way, technology impacts the post-military life of the veteran today. We have a platform that amplifies and extends our voice, but it also amplifies and extends our discontent. For every story of a veteran knocking it out of the park, there are stories of veterans who are struggling: the raising suicide rate. The challenge in accessing mental health services. For every image of the veteran returning to their community and rolling up their sleeves in order to get to work, there is an image of the disgruntled veteran who just wants to be left alone and shake their fist at the sky. The successful veteran balks at the image of the struggling veteran, and the struggling veteran is angered by the attention given to the successful veteran.

Each Side of the Divide has a Responsibility

The Successful Veteran, developing into the Greatest Generation, has the responsibility of being aware that a large portion of it's brothers and sisters are being left behind. As the lead of the cohort outdistances those that follow, there is a need to circle the formation back around and support those that are not as successful. This is happening through organizations like Team Rubicon, The Mission Continues, Team Red, White and Blue. Organizations like the Warrior Wellness Alliance, established by the George W. Bush Presidential Center, are working to connect these organizations in a comprehensive way.

The Struggling Veteran, on the other hand, has a responsibility to themselves to take advantage of the resources that are available to them. To not stay stuck in the Lost Generation, but to improve themselves and accept their role in the Greatest Generation. This is often a conscious choice, and has much to do with mindset. Again, as written before, mindset and mental health and wellness are the key to whether or not a veteran is going to succeed or struggle in their post-military life.

Transition Between Both Sides of the Divide is Possible

A veteran can start out struggling, and ultimately succeed. They can be traveling down the path of being lost, and finally get it together and become great. I've seen it personally in my work as a mental health counselor; the veteran is struggling with some different aspects of veteran mental health, and experience some challenges. Involvement in the criminal justice system. Homelessness. Addiction. All of the indicators of underlying, untreated mental health challenges. And, after setting things right, they get back on track, course-correct, and begin to succeed.

Conversely, the successful veteran is only one significant event away from becoming lost. When I was a program director for a homeless veterans program, I often said that there was not much of a difference between the guy running the place and the guys and gals living in the place. I am only one bad decision away from being a participant in my local Veteran court, rather than a member of the team of it. I work diligently not to make those bad decisions, but the possibility is there.

If we had a choice, we would choose success and a peaceful post-military life. What can we do to help those veterans who are struggling to be aware that they can choose the same?

PART 3

Who We Are: Transformation and Change

I will never be a civilian. I was, once; I became a Soldier when I joined the military. Now that I've left the military, I have become a third thing, a mix of soldier and civilian: this thing called a Veteran. I will be a veteran at least twice as long as I was a soldier (God willing) but it will always be a significant part of my identity.

Change is difficult; it's uncomfortable to leave the familiar and try something new. The change from civilian to soldier was hard and demanding...but it was also required. It was a complete transformation: physically, behaviorally, psychologically, and emotionally. The change from soldier to veteran was demanding too, but for many of us, the transformation is not complete. We change physically (unfortunately) as we stop doing physical training, but also the other extrinsic factors of the military go away. We don't wear the uniform anymore, we can walk outside without our headgear on, those physical manifestations of the military are gone.

We also change behaviorally, to some extent. We don't march everywhere we go, we don't stand outside in the freezing rain for seemingly no reason. If we do continue to engage with firearms, it's as a hobby or for sport, not as a part of our job. Some behavior carries over; as of the writing of this book, it's been over five years since I retired, and I still get up before the sun comes up. Sleeping in, for me, is 5am. And I feel like I've lost half the day.

What some of us don't do is change psychologically or emotionally. It's not required; there's not some post-military drill sergeant demanding that we change the way we think and feel. Sometimes, we don't even know that it's necessary. It's only when it gets in the way and we find ourselves stuck in post-military life that we have to figure out how to make the transformation complete.

Veterans: Who You Are Is Not Who You Were...

...and who you are is not who you're going to be. When a veteran looks at themselves in a mirror, it's a toss-up about who they're going to see: who they were, the military service member? Who they are, just them as they are right now? Or them in the future, a competent professional? When I was in my primary leadership development course back in the '90s, one of the instructors told us a key principle of leadership: "We are three people. Who we think we are, who others see us as, and how we really are." For veterans in post-military life, these three perceptions can be wildly different.

We all have different personas, different roles that we play in our lives. If you sit for a minute and think of all the roles you play, you can probably easily come up with ten or more. Veteran, if you served; spouse, parent, cousin, child, grandchild. For me, therapist; writer; podcaster. Friend, enemy (although hopefully not too much of that). The challenge for many veterans I work with is that they want to be who they were, rather than who they are. These different roles don't agree with each other; I am veteran, but I want to be service member. The farther away each of these personas are from each other, the more challenging life can be.

When I talk to clients about this, I often explore these three personas, and I've found that we generally carry three separate ones: the service member/veteran, the person we present to the world, and who we really are. To illustrate this concept, I'll take a cue from a podcast recently discovered about the psychology of Batman, The Arkham Sessions[1]. One of the co-hosts, Dr. Andrea Letamendi, describes the three personas in this way: there is Batman, the vigilante; Bruce Wayne, the playboy millionaire that is presented to the world; and there is Bruce, the real guy underneath both.

For veterans, I see this in a similar way. There is Sergeant First Class France, the military me, morphed into my identity as a Veteran. Then there is Mr. France, the post-military professional, who does all the meetings and networking and shaking hands and business cards and stuff like that. Then there's Duane, the guy underneath both of them. The problem arises when Duane identifies so heavily with the Veteran persona and is unable to transition to the professional persona.

Sometimes, the Veteran Wants To Be Who They Were

This is when the veteran longs for the glory days[2] of the past. They wish they never got out of the military. They want to be back where they were comfortable, where things made sense. If my identity is wrapped up in being a high school football player, then I never move on from it. If my identity is wrapped up in the service member I was, then I'm stuck in that

76

persona. I relive my glory days, living in the past, becoming more disillusioned as time goes on, more frustrated because who I was is not who I am.

Sometimes, the Veteran Doesn't Want to be Who They Are

This can compound the problem of wanting to be who they were. Now, they're stuck. They're racked with guilt about the things that they did in combat, if they deployed, or angry about how they're treated upon their return. The real them behind the veteran persona, the one alone behind closed doors, is depressed, bitter, afraid, lonely. Unsatisfied with their current life, feeling unloved and unproductive. This can increase the pull of the military persona; we relive the days where we were not depressed, afraid, and unproductive. It makes us long even more for the time when we mattered, because we don't feel like we matter now.

Sometimes, the Veteran Doesn't Know How to Become Who They're Going to Be

So if Sergeant France is who I really want to be, and Duane is someone I don't like, it's going to be hard to develop into Mr. France. The professional persona, the post-military persona, becomes unobtainable. We don't know that that person looks like, and we don't want to know. We want to be back in the service, we want to be comfortable. In some ways, we want to be stuck, even though it feels like crap.

The challenge is, some veterans don't realize that they have so much more to give to the world. Many believe that their commitment to serve did not end when they took off the uniform, but their ability to do so did. As long as we believe that to be true, it is true; if we think that the best days of our lives are behind us, when we have decades of life in front of us, then hope is not on the horizon.

Balancing These Personas is Key to a Satisfied Post-Military Life

The more we want something, but can't have it, the more upset we become. Reality is the fact that I'm no longer the 25 year old paratrooper. I'm no longer the leader I was in the military, the Platoon Sergeant or First Sergeant. Wanting to be that, and not being able to, is like me being jealous of my former self, and that's a losing battle. Instead, appreciating who I was, and incorporating that into who I am, which helps me develop into who I'm going to be, is a more balanced approach, and comes with much less distress.

Once we find balance between our military persona, our professional persona, and who we really are, post-military life becomes a whole lot easier.

Awareness without Action Does Not Lead to Change

You don't climb a mountain by saying you wish you could. One of the challenges that I see with the veterans I work with as a mental health counselor is an experience of being stuck. Regardless of how things were when they were in the military, they seem to feel bogged down in their post military life. How often do we say that we want things to change, wish they were different, but still do the same old thing?

One of the things that I always stress with my clients is the need to become self-aware. What am I thinking. What am I feeling. How is the environment affecting me, and how am I impacting the environment. Without this awareness, we may know that there's a problem, but we don't know what it is. And if we can't define the problem properly, then we can't come up with a solution.

There are a lot of things that get in the way of taking action. In applying the theory of planned behavior[3] to veterans seeking health care, the authors identified a number of barriers[4] that keep a veteran from taking action. These include stigma against seeking help, the beliefs of others about help seeking, and not knowing how to access support. Without overcoming these barriers, change won't happen.

Awareness Is Not Enough

I've said before, and I say it often to veterans I work with, that awareness is the key to recovery when it comes to veteran mental health. Often, veterans don't know that things are different unless it's brought to their attention. A mentor of mine, Dr. Steven Kidd, says it this way: "people will seek mental health treatment because of one of three things: their lawyer, their lover, or their liver." Something in the world brings the problem to the veteran's attention; but simply being aware of that is not enough.

In a study released in January of 2018, the National Academies of Science, Engineering and Medicine interviewed over 4000 post-9/11 veterans, and found that over half of them screened positive for some form of mental health condition. Of the number that screened positive, 16% of them knew that there was a problem...but refused to engage in treatment. Knowing there is a problem is not enough to solve the problem, it's only the first step.

Negative Thinking Holds Me Back

Another thing that keeps us from taking action is that we're caught in a cage of our own construction. We tell ourselves that something is impossible, when it really isn't. That we "can't" for whatever reason. We wouldn't be allowed, we don't have the ability, the task is impossible. Going back to the quote at the beginning of this article; we may stand at the foot of a mountain and look at the top and want to get there, but if we tell ourselves we can't do it, then our desires will not be fulfilled. Wanting to be on the top isn't going to get us there; we literally have to take steps to make it happen, in this case. This speaks to the Audacity Principle; if I put limits on my own sense of what is possible, then I'm not achieving all that I can. Or even all that I want.

My Perceived Limitations are Greater Than They Are

This is something that I see with veterans. We allow conditions to define us, instead of moving forward in spite of the conditions we find ourselves in. "I had to get out of there because of my anxiety" or "I can't get going because of my depression." Action is moving forward *in spite of* instead of not moving *because of.* It's as if our anxiety or depression or disability, if we want to call it that, is an external force that we have no control over. We *do* have control over it; I am certain of it, because I see the evidence around me every day. If we accept our limitations as the upper limits, rather than acknowledging them as simply obstacles to navigate, then we will remain stuck. We will not take action.

It's Easier to Want than to Do

Change takes effort. Whether it's physical health, psychological health, spiritual health, it all requires work. I often hear from veterans; "I wish this would just go away overnight." It didn't happen overnight, did it? Why would we think that we could instantly untie a knot that has been forming for years? The small steps…action…are what will eventually lead to a big change.

It was once described to me in a crude way, but I haven't heard a better one. It's as if we're sitting in a warm pool of our own crap. We don't necessarily like it, but it's known and it's comfortable. Getting out of the pool takes effort and work; and that effort is uncomfortable. We start to think of those barriers: "there's no help out there." "Nobody really gives a crap." "Nothing I will do will make a difference anyway." Then, we slide back into our known and comfortable pool of crap, and say, "I wish I could get out of this."

A Little Effort Goes a Long Way

Start something today. Make a small change. If you want to write, don't start on a novel; start with a paragraph. If you want to lose fifty pounds, start with a walk around the block. Small actions, done consistently, will accumulate into large results. You just have to take action.

How We View The World Impacts Veteran Mental Health

This may come as no surprise, but how we see the world can be helpful or harmful to mental health and wellness. This isn't just about positivity or optimism; studies show that our outlook on the world and other people can protect us from the impact of negative life events. In my experience, this is especially true for military service members and veterans. A positive, even if unrealistic, world view can protect us from the impact of negative experiences. Similarly, a negative world view can make a bad situation worse.

Unless we look closely about how we see the world, we're not often aware about our assumptions. Awareness can then lead to understanding, and deliberate change if the way we see the world is getting in our way.

Assumptions About the World Impact Reaction to Stress

Sometimes, a service member will go into a situation with negative assumptions about the world. We go into the military carrying whatever experience we have with us, good or bad. Adverse childhood experiences...abuse in multiple forms, exposure to domestic violence, substance abuse in the household, disrupted families, family member incarceration...has been shown to[5] increase the risk of trauma-exposed service members to traumatic stress reaction after the military. These adverse experiences give us an explanation about the world. Other people can't be trusted. We have no control over our own environment. Bad stuff happens randomly and without observable cause. If we come into a situation with negative assumptions about the world, then we are at greater risk to react negatively to stressful events.

At other times, a traumatic or stressful event will cause a service member to develop negative assumptions about the world. If we have certain positive assumptions about the world, these assumptions may be shaken or even shattered when we experience stress or exposure to traumatic situations. In a study looking at the impact[6] of world assumptions and mental health and wellness, the results indicate that the way we see the world can be challenged in the face of trauma induced by other people. In other words, we think the best of people...until we're proven wrong in the worst way.

Attributing Survival to Luck rather than Skill

The military is an inherently dangerous occupation, whether a service member experiences combat or not. Any number of things could go wrong on a daily basis. Multi-ton vehicles, live ammunition, deliberately stressful situations, dangerous activity; all of these things have the potential to go horribly wrong. One difference that I've noticed about veterans that I work with is that some attribute their survival of dangerous situations to skill, and others attribute their survival to luck. When it comes to skill, either that of the service members of others, it implies a measure of control over a dangerous situation. "We made it out of there because we knew what to do" depicts a measure of confidence in our own abilities or that of our team.

Contrast that to someone who says, "I shouldn't have made it out of there alive. I had something looking out for me." Survival is random. Out of our control. We made it out because of luck, or fate. Many veterans have "the day I nearly died" stories; not just survival of a direct attack or improvised explosive device. "Almost died" could mean that we got somewhere ten minutes earlier, or ten minutes later. Once, the vehicle I was in nearly fell off of a six-story cliff; disaster was averted because of a one-foot brick wall that caught our axle before going over. Had the wall not been there, I wouldn't be here. And believing in luck isn't always a bad thing; but when we explain survival only to luck, then it goes back to the unpredictability and lack of control of the adverse childhood experiences.

Shifting Point of View from Negative to Positive

This is where it gets difficult, but it's also easy. Many times, we're not aware of our negative our assumptions about the world. When that awareness is developed, we can then decide to do something to change it. Sometimes, it's as easy as changing what we think to ourselves. "I shouldn't have made it out" can be changed to "obviously I should have made it out, because I did." It's recognizing the negative way we're seeing the world and finding evidence to the contrary. Finding good things in the world to counteract our belief that the world is not benevolent. Finding evidence of people held accountable for their actions in order to restore a sense of justice.

I hear you already, of course. "Easy for you to say, big guy. You don't see how much the world sucks." I can tell you that it's not easy for me to say; my basic outlook is pessimistic, and I have worked hard over the years to develop optimism. And I can also tell you that, just as you say that I don't see how much the world sucks, you don't see how great the world is. We each have our blinders on in a certain way, and can certainly change how we see the world[7]. I know, because I've done it. I've seen others do it.

After we become aware of how our negative viewpoint is impacting us, it's up to us to change. If we want to.

Veterans: Strangers in a Strange Land?

" I don't belong here."

"This feels weird."

"I feel like an alien."

These are all things I've heard veterans say after leaving the military or returning from combat. Something changed, and they're not sure what. It was them, sure, but it was also the community that they came back to. For many veterans I talk to, they feel like strangers in a strange land. They seem to be visitors from somewhere else trying to make their way in a world where they don't understand other people, they can't figure out the rules, and sometimes don't really want to.

The phrase "stranger in a strange land" originally comes from the book of Exodus: Exodus 2:22. Many know the story of Moses; raised by Pharaoh's daughter, he fled Egypt for killing an Egyptian who was beating a Hebrew slave. In exile from Egypt, he named his firstborn son "Gershom," which means, "Stranger in a Strange Land." Moses went from a culture that he grew up in, one that he knew the rules and understood how things were done, to the wilderness. No guide, no easy transition, just one day he was surrounded by familiarity and the next he was out in the wild. Sound familiar? How shocking would it be to be taken from everything you know, everything you're comfortable with, and be dropped into a place that you know nothing about? Many…not all, but enough…veterans feel this way upon returning back to their communities.

Why is that? Why do some former service members make a smooth transition, while others experience challenges? Why do some seem to be able to fit back into their community, while others feel a tangible gulf between themselves and others? Much of it has to do with not understanding how important mental health is to the transition process. Another large part of it, in my opinion, is a lack of awareness on the part of the veteran: awareness that change needs to be made, awareness that they themselves have changed, and even awareness that they have the ability to make the transition back.

Leaving the Military is Inherently Stressful

I know it is. I've done it. I've talked about it and wrote about it. How I'm going to pay the bills and make a living are only part of it; understanding how to operate in non-military settings was also a challenge. A study published in 2017[8] argues that PTSD is not the primary source of discomfort for returning service members; it's transition stress. Transition stress, among other things, is the subject we're talking about; a veteran not

82

feeling like they belong in their community. In this article, the authors address such subjects as the loss of the military self, service-connected nostalgia, moral injury, and the civilian-military divide. Each of these demonstrate the concept that a veteran feels separated from the very society that they served to protect. Being in one place, and longing to be in another. Existing in one time, and desiring to be back at a place that made sense.

It Just Feels Weird to be Someplace Unfamiliar

I experienced this on a small scale, as have many other combat veterans. My base in Iraq, Camp Rustamiyah, was small, about a square mile or so. Big, compared to other combat outposts and smaller bases, but minuscule compared to giant military installations like Bagram or Talil. Several times during that deployment, I had an opportunity to leave my small FOB and go to these larger bases...I felt like a kid leaving a small town in the middle of Missouri and being dropped into the middle of New York City. The size, the activity, the sheer unfamiliarity of it all...it was overwhelming. I found myself wanting to get back to my small camp, where I knew where everything was, and everyone knew me...where things made sense.

Learning how NOT to be a Stranger Takes Effort

We feel like a stranger in a new place until we get comfortable there. My first duty station was Germany; new language, new culture, I was literally a foreigner. There were two things that I could have done. First, I could have kept myself separated from German culture, not tried to learn the language, and stay in the barracks...continued to be a stranger. The second thing was that I could have ventured out from what I found to be familiar, and learned about the country, the language, the people. I knew people who did the first one, and they didn't have that great of an experience; they continued to be a stranger in a strange land. Others, admittedly, "went native" and immersed themselves in the culture *completely*. There's nothing wrong with either of those, if they didn't bother the person or impact their family.

The same way we get used to a new place is the same way a veteran stops being a stranger in their own house, their own community. We adapt, as we adapted when we were in the service. We learn the new language, embrace the new way of doing things. We deliberately choose to not set ourselves apart from others, and instead learn to integrate what we know with what they know.

By stepping into the gap, the civilian-military divide is reduced, and the veteran stops feeling like a stranger in the land that they love.

Prepare for the Storm While the Sky is Blue

One of the key aspects of military service is training. Training, training, training. We train in the heat, we train in the rain, we train on the water and in the sky. There's never a question about why we do it; we know why we do it. Preparation.

I am often reminded of one of Aesop's fables, the Ants and the Grasshopper. In the story, the ant works all day brining food back to the anthill, while the Grasshopper does nothing; he even laughs at the ants for working so hard. Of course, when winter comes…when the storm comes…the grasshopper has nothing, and the hard work of the ants pays off.

So what does this have to do with veteran mental health? We hear it, an ounce of prevention is worth a pound of cure, a gallon of sweat in training saves a pint of blood in combat, but how does this apply to mental health in post-military life?

A Lack of Preparation Can Make a Bad Situation Worse

If I'm not working on my mental health and wellness, if I haven't developed awareness or recognized that my emotions have changed, then when stress comes, the impact will be even greater. A veteran goes through a breakup: sure, that's bad. If that same veteran has not learned to understand and control their emotions, or is dealing with unresolved trauma, then it goes from bad to worse. If we're trapped in negative thinking patterns, then we become caught in a cage of our own construction. This is part of the downward spiral; if we don't get our head space and timing set right, then our lack of preparation can make a tough situation into a crisis.

A Lack of Preparation Might Indicate Denial

Like the grasshopper in the story, it might be that our lack of preparation could be the result of denial that the danger is real, or that it applies to us. It's like being blindsided, caught flatfooted. Sure, there is a suicide epidemic among veterans, but that will never happen to *me*. My buddies have a drinking problem…*I* don't. We don't start thinking about understanding how combat and the military changed us until *after* a crisis, because we don't think it did change us. Or, if it did, what did you expect? By not preparing for a potential crisis, we could be saying to ourselves, "that may apply to everyone else, but not to me." Sound familiar?

A Lack of Preparation Might Indicate a Lack of Foresight

Instead of denial, a lack of preparation when it comes to mental health could indicate a lack of foresight. We trained in the military because our leadership anticipated the situations we would likely face in the future. We had large scale training exercises for things that would happen months or years from now. Before we went on missions, we ran through rehearsals to prepare for things that would happen minutes or hours from now. It was based on an anticipated possible future, and we plan for all contingencies.

If we don't start understanding how our mental health impacts our post military life, we might be demonstrating the lack of foresight that helped us in the military. Without understanding the potential danger, we may think that the sky will always be blue, that we will never face a crisis moment. How realistic is that? It's not a denial that it can't happen to us, but a denial that it can't happen at all. Equally dangerous.

A Lack of Preparation Might Indicate Capitulation

Another potential reason for not getting our mental health straight is one of the prohibited words in the military: quitting. Giving up. Accepting that the situation we're in is never going to change. In the recent National Academies study of the VA Mental Health System, they identified that more than half of the veterans who have a mental health need don't perceive that they have a need. Only three out of ten veterans who have a need seek services; that means that approximately there are about 16% of veterans who have a need for mental health treatment, know they have a need, but yet won't get the help.

Sure, there are reasons for this. A lack of quality care, both inside and outside the VA. Other barriers, like time, distance, money. Perceived stigma. When did we allow obstacles to stop us in the military, though? If we stop preparing, then we stop caring. When we stop caring, we're on a slippery slope, and when a storm does come, we just don't give a crap about it. Dangerous place to be.

Preparation Comes in Many Forms

Okay, now what? So how do we prepare? This is a good start. You're taking the time to read something that could spark a thought, and might find something more that catches your attention. There are tons of online resources. Check out some of the content that the VA has put together to combat stigma[9]. Take a short course from PsychArmor[10] to raise awareness about the impact of the military on mental health and wellness.

Character is Revealed in Crisis

When the world falls apart, who we really are is revealed. Our preparation, or lack of it, shows itself. If we weren't proactive with our physical fitness, it shows when the PT test comes. If we half stepped our way through training, it comes out. The same thing will happen if we don't prepare for a potential mental health crisis. Dr. Martin Luther King says, "The ultimate measure of a man is not where he stands in moments of comfort and convenience, but where he stands at times of challenge and controversy." In other words, when the sky is blue, we're not tested; it's what happens in the storm that is the true measure of our worth. On that day, the work that we put in is revealed.

Isolation, Solitude, and Veteran Mental Health

We don't function well as human beings when we're in isolation – Robert Zemeckis

Humans are social creatures and pack animals. Our basic nature needs interaction with others to satisfy our biological needs, our safety needs, our social needs. We need other people to survive and thrive. One challenge with veteran mental health and wellness, however, is that it often drives isolation rather than connection. This complicates recovery significantly. This problem doesn't just apply to the veteran suicide epidemic, but it's a good illustration of the point. Dr. Thomas Joiner is a recognized expert in suicide research, and his Interpersonal-Psychological Theory of Suicidal Behavior[11] illustrates the dangers of isolation. Dr. Joiner tells us that the evidence shows that there are three factors that are involved in someone acting on suicidal thoughts. These three factors are a sense of burdensomeness, social alienation, and the acquired ability to enact lethal self-injury. In other words, if a veteran feels like their problems are too much of a burden for those around them, they are isolated (either physical or psychologically) and have the ability to hurt themselves, then suicide is a significant likelihood. When it comes to isolation, however, it can be complicated.

Isolation Is a Risk Factor for Increased Psychological Suffering

It has been widely known that social isolation is a key risk factor in challenges with mental health and wellness. Withdrawing from social relationships has been shown to be a barrier to care and a factor becomes a barrier to care, according to one study[12] in 2009 by Dr. Beth Cohen and her colleagues. It is also a factor that impacts military families; one study conducted by Dr. Sean Phalen and his colleagues, published in 2009[13], found that social isolation was a significant problem among caregivers of veterans with traumatic brain injury. These findings aren't new, either; a study looking at caregivers of Vietnam Veterans with PTSD[14], published in 1993 by Dr. Kathleen Jordan and colleagues, found that social isolation among caregivers was one of the most commonly reported problems. Withdrawing from others…family, friends, social networks…increases the sense of burdensomeness described above.

Isolation as Avoidance

One of the problems with isolation, however, is that sometimes it feels good. One of the key aspects of Posttraumatic Stress Disorder is avoidance of stimuli that reminds the veteran of traumatic events. Not all veterans have PTSD, of course. Someone who had been caught in a crowded market, or someone who had repeatedly been exposed to wide open spaces where danger was a constant threat, would naturally want to NOT be in those situations. We will avoid that which makes us uncomfortable. I have spoken to veterans who used to enjoy going to huge music festivals before they joined the military, but couldn't even stand the thought of going to a small club concert after their career was over. In this way, isolation became a way of avoiding unpleasant physical and emotional responses. Isolation also serves to help veterans avoid the stigma of mental health concerns. If someone constantly hears, "you've changed since you got out" or "why can't you go back to being the way you were," then these comments will add to that sense of burdensomeness. "What I've become is unacceptable to those around me" is the thought that we tell ourselves. Since we don't want to be rejected, and since we don't want to bother those around us, we start to pull away from others. This, in turn, leads to an echo chamber in which the veteran only hears what they think, and no one is around to tell them otherwise. Isolation, like other forms of less than appropriate coping techniques like alcohol and other drugs, works well in the short term…but can be extremely harmful in the long term.

Isolation Versus Solitude

Other times, however, being alone is truly beneficial. There is a difference between isolation and solitude. In an article published in 2015, Dr. Jacob Stein and Dr. Rivka Tuval-Mashiach differentiate between the two states[15]. Social isolation is associated with loneliness, while solitude is not. If someone is socially isolated, they want to be alone (in some cases) but also truly do miss interaction with others; they want it and don't want it at the same time. Solitude, however, is a state of desired aloneness; there is no loneliness associated with it. Sometimes this desire for solitude is an acquired trait. After a veteran's experiences in the military, they may desire less interaction with others; we're not avoiding the crowd because it reminds us of Baghdad. We're avoiding the crowd because being part of the crowd no longer interests us. As I often tell the veterans I work with: if you know why you're doing something, and you're not hurting yourself, your family, anyone else, or breaking the law, then so be it. Building a compound in the mountains out of anxiety? A desire to avoid anyone and everyone? Not helpful. If you buy some land and build a house for you and your family simply because you enjoy the peace and quiet; helpful. Alternatively, a desire for solitude may just be part of your nature. I'm a naturally private person; although I do somewhat live a private life in public[16], I'm an introvert with extroverted tendencies. I had, and continue to have, a small group of very close friends, and am content with my wife and kids. That's different than being isolated; it's being comfortable with aloneness.

Value Solitude, Avoid Isolation

Whether you're experiencing solitude or isolation, the key is understanding which it is. How is it impacting you? Are you isolating in order to avoid unpleasant interactions, but really do want to interact with others? Then it's likely isolation, and too much of that can be dangerous. If you're struggling with isolation, the best thing to do is to avoid avoiding[17] and reach out to talk to someone. It's not as painful as you imagine.

Accepting Reality and Changing If to Is

Afghanistan, 2010. Our unit had been in Regional Command East for a year, and were preparing to leave; as a platoon sergeant, my platoon leader and I had the responsibility for making sure the new folks coming in were ready to take over the mission. Known to us as a "RIP," short for "Relief in Place," this can sometimes go well and sometimes go poorly. Things were going well this time, but there was one problem. One of the incoming leaders had a habit of shrugging his shoulders and using a well-known phrase in the military: "It is what it is."

It got on our nerves. The phrase, "it is what it is," can have multiple meanings. For this particular leader, the meaning was, "this is the way that it is, there's nothing I can do to change it, so I'm not going to even bother." It's giving up, capitulation. A form of learned helplessness and even laziness.

The problem, though, is that reality really is what it is, not what we want it to be. Another habit of service members and veterans is that we often use the word "if" when we should really be using the word "is." We look at the past from an "what if" point of view, and look at the future with an "if only" point of view. What if I had decided to turn right instead of turning left on that route. What if I had spoken up when I felt something right. And for the future: if only I had a better job. If only I could get out of this town. If only my (insert individual here) were more supportive, life would be great.

Life don't work that way.

The What If Game Denies What Is and What Was

A favorite phrase of mine from Irvin Yalom comes from his book, Staring at the Sun: Overcoming the Terror of Death[18]. In that book, one character says of another, "sooner or later she had to give up the hope for a better past." My question often to veterans that I work with is: how long are we going to wish for a better past? Our past is what it is, and we can frustrate ourselves by wishing it was different. The reality is that I grew up in St. Louis, MO and my first roommate in the Army grew up in Hilo, HI. What if I had grown up in Hawaii? Would I be surfing right now instead of writing? Who knows! And while it may be an interesting diversion to think about how things would be different, this isn't the comics and there are no alternate universes. Our past is our past.

Changing If to Is in this context is accepting the past for what it was. Regardless of looking back on it and wanting the past to be different, it can't be and will never be. Not accepting the past for what it is can lead to regret. Once thought to be a uniquely human emotion, a study in 2015[19] shows that rats exhibit the neurological reaction in regret in decision making. Regardless, regret is something that we humans are intimately familiar with; in the linked article, authors Steiner and Redish define regret as our recognition that choices made in the past resulted in worse outcomes than a different choice would have. Regret also influence future behavior; current mood, of course, but also the memory of a "bad choice" in the past can cause us to second guess our choices in the future. Accepting this reality releases us from the impact of regret.

The If Only Game Denies What Is Happening Now

Looking at the future, wanting our circumstances to be different from what they are also gets in our way. When I hear a veteran say, "if only I had another job," my question to them is: are you looking? Are you taking advantage of the different opportunities available to find something different? "If only" rejects our current reality for a future fantasy. This can apply retroactively, of course; "If only I hadn't been stationed at Fort Polk" then life would be different. The If game of wishing for a different reality, either past or future, can get us stuck in wanting something to be different and disappointed that it's not.

Changing if to is in this context is figuring out how to move forward from the place we are, not from the place we want to be. It's all well and good to say that things would be easier if the situation were different, but the situation isn't different. Wishing it was makes things more difficult.

Accepting Reality to Progress, Not Give Up

So back to "It is what it is." Accepting reality and figuring out how to make changes is different than accepting reality and thinking we can do nothing about it. If we find ourselves lying in the middle of the road, the reality is that we're there. Thinking "it is what it is" and deciding that there's nothing we can do about it means that we're going to get run over. Doesn't make much sense, right? Instead of using "it is what it is" as a reason to not try, we need to use it as a reason to try more.

Accepting the reality of our past, current situation, and future potential can be liberating, not limiting. We just need to decide to change If to Is.

Veterans, Do We Accept Limitations or Acknowledge Limitations?

"It is the same with people as it is with riding a bike. Only when moving can one comfortably maintain one's balance." – Albert Einstein

Imagine getting to a point in your life when you say, "Well…this is it. Life can't get any better (or worse) than it is right now. I guess I better just tie a knot below me and above me and hang on with all I've got." I don't know about you, but for me, that's pretty stagnant. Comfortable, maybe? Complacent, there's that word that veterans hate so much. Fatalistic, capitulating, settling…stuck. Never moving forward, never growing, never experiencing what it's like just a little bit higher up on the rope. Maybe, sure, there is something going on that limits or impedes growth or movement. It is up to each of us, however, to either accept that limitation, or acknowledge it.

I have seen veterans who believe we should just accept our limitations when it comes to mental health. I've heard people who support them say the same thing. I've had people ask me: if a veteran has PTSD, are they stuck that way? Is a veteran with severe PTSD forever doomed to a life of isolation, startle response, avoiding crowds, angry outbursts that are explained and accounted for because, "well, of course they act that way…they have PTSD?"

The answer is yes, if the veteran and those around them believe that they should accept their limitations and not attempt to push beyond them. At the same time, the answer is a resounding and emphatic no, if the veteran believes that they should acknowledge their limitations, and thrive in spite of them, not stagnate because of them.

Fixed Mindset and Growth Mindset

We limit ourselves by accepting that which limits us. Carol Dweck, currently a professor at Stanford University, has conducted extensive research regarding mindset and viewpoint[20], and has developed the concept of a growth mindset versus a fixed mindset. Dweck's work focuses primarily on how people see their ingrained traits, such as intelligence and personality. If someone sees these traits as fixed, unchanging, then they will not attempt to change them, will not attempt to grow. This is simply accepting that we have a finite amount of whatever is in us…brains, talent, guts, compassion…and we should just ramble around the world making the best of what we've been dealt. These are our cards, can't change them.

Compare that to Dweck's concept of someone with a growth mindset. We begin with what we have, but we also have the ability to cultivate those attributes into something more. Talent can be developed. Compassion can be learned, guts can be discovered, we have the ability to increase what we have and develop what we don't. By acknowledging where we're at, but not accepting it as the final destination, we have the ability to change and grow.

Accepting Limitations versus Acknowledging Limitations

So how does this relate to veteran mental health and PTSD? The fixed mindset is one that simply accepts that there will be no change. In a recent conversation, a connection described a veteran she met as someone who was "pretty clear no one should expect anything from him." That veteran had gone as far as he could, it's not going to change, he (and those around him) have accepted his PTSD as a limiting condition and have made it one.

Compare that to the same connection, who saw another veteran with a growth mindset who struggled with PTSD and anxiety, but "he's at a big event, leading a team. Anxiety visible but he's out there." This, then, is a veteran who is acknowledging how his condition impacts him, but chooses not to accept that it limits him. He grows and thrives in spite of his limitations, not becomes stuck because of them.

Accepting Limitations Keeps Us from Change

Change is hard. It's uncomfortable, certainly much less comfortable than staying how we are. If we are not constantly striving to improve ourselves, then we are not growing. Why stay stuck in a place where we are unhappy, miserable, and make life challenging for those around us? There has to be a good reason for it, and usually it's because we're in a fixed mindset. What we have is what we have. We can't change it, we can't grow from it, we can't learn from it.

On the other hand, change is good. The concept of Posttraumatic Growth[21] is a result of a growth mindset, compared to a fixed mindset. When we get knocked to the mat, we can stay there. Or we can get up and get back in the fight. If a veteran who is struggling with anxiety or depression does not think they can change, then they will remain in a painful, unpleasant experience for a very long time. Maybe forever.

Acknowledging Limitations Means Change is Possible

I have seen it before, I see it nearly every day. The veteran who struggled with ten years of addiction, but is now has eighteen months of sobriety. The veteran who could not bear to think of the trauma that they endured, much less talk about it, who is finally opening up to a trusted source after years of pain. A veteran who had such a traumatic brain injury that they are quite literally a different person from before the event, but is building a new life for themselves with what they have, rather than mourning the life they no longer have. The veteran who says, "I'm tired of this, tired of it impacting my family, I'm just not going to let it get to me anymore."

This is more than just making lemons out of lemonade, more than just the "suck it up and drive on." Instead, this is taking a hard look at the realities of life and what's going on in our mind…and then choosing to build a life worth living, rather than choosing a simple existence.

And that is going to lead to the enjoyable post-military life that we all desire and deserve.

Protecting Your Core Self from the Arrows of Malevolence

The arrows of malevolence, however barbed and well pointed, can never reach the most vulnerable part of me – George Washington

During George Washington's second term in office, he was getting bombarded by criticism. Federalists, who wanted a strong central government, were dissatisfied with him, and ambitious Republicans, who advocated for more power to the states and the citizens, were both at his heels. (you can read more about the challenges he was having here[22]). Having fought and won a war, liberated a nation, and helped create a government, he was quite simply tired and done with it all.

Sound familiar? If ol' George was hit from all sides with criticism and found it hard to deal with, you know that it's true for us today. Many of the veterans I work with struggled with toxic environments when they were in the military; I dealt with it myself. Gossip, backstabbing, leaders who use their influence for personal gain. Peers who want to advance in rank, subordinates who want to set you up and take your place. Humans can be downright nasty to each other; and that's to people who are on the same side as us.

When a service member leave the military, if we don't adequately protect the most vulnerable parts of ourselves from the arrows of malevolence, then there are going to be some problems in our post-military lives. And these arrows don't just come from others; often, we become our own worst enemy by internalizing these attacks and turning them in on ourselves. Realizing that we're doing it, and protecting ourselves, is key to a more peaceful life.

The Arrows of Others

It's gonna happen. The boss. The romantic partner. The random person in the store who's having a bad day. Someone is going to struggle with something and take it out on you, just like you're going to take it out on others. It's not okay, of course, but it happens. It might be displaced aggression. They're really angry at someone or something else and taking it out on you. Or they might be doing or saying something deliberately to hurt you. The key point to remember when dealing with the arrows of malevolence that come from others is that it's not about you.

It may be hard to realize that, when you're getting snide remarks from your coworkers or passive aggressive comments from the people that work for you. It sure feels personal. Any attack is personal…but it's not always personally about you. A colleague of mine often says, "half of the people in the world don't care that you have a problem, and the other half are glad you have one." Maybe they're not glad that you're suffering, but it may be that your suffering may mean that they're not. Dealing with these sharp elbows isn't always easy, but life's going to throw them at us.

The Arrows from Ourselves

The more painful and dangerous arrows of malevolence often come from ourselves. In a recent conversation[23] with Dr. Jannell MacAulay, she talked about how she was encouraged by her father from a young age. She was told that she would grow up to be a fighter pilot or a submarine commander...neither career fields which were open to women at the time. Her father spoke words of encouragement into her, and she internalized them (and became the pilot, by the way). If the words that were spoken into her at a young age were not encouraging, however, but malevolent...they would have been internalized in the same way.

If a service member's experience in the military was one that they were repeatedly told that they were worthless, or lazy, or stupid, then they will start to believe that themselves. They will use the same words to describe themselves, and the arrows of malevolence are no longer attacking from outside the walls, but inside. "You can't" becomes "I can't" and "you're a waste" becomes "I'm a waste."

Often, we don't even have to be told these things, but interpret them. Leaving the military can be hard, but it's even harder when it's unexpected. I've had veterans who were medically discharged say that they felt like they were thrown away like trash. They interpret that to mean that they're worthless and rejects, and beat themselves up accordingly.

Protecting our Core Self

The key to protecting ourselves from both of these sets of arrows is to develop a core sense of self that is not vulnerable to attack. Use whatever metaphor you'd like; a shield to block the arrows, or steel walls to stop the bullets. Psychologically, building or reinforcing a core set of beliefs about yourself that are healthy and protective. Realize that, when you're experiencing the malevolent arrows of others, it's not about you; it's about them. Their problem does not need to become your problem. Also be aware that first, you're beating yourself up, and second, that you don't have to.

The damage that we do to ourselves is sometimes greater than the damage from others. When we build a cage of our own construction[24] with the words can't, or shouldn't, then we're not doing ourselves any good. We have to assert, like President Washington did, that these arrows of malevolence can never reach the most vulnerable parts of ourselves.

For Every Mile of Darkness, We Need an Equal Distance of Light

Hope is being able to see that there is light despite all of the darkness- Desmond Tutu

Veterans can be a pessimistic group. I'm not saying that there's something wrong with that; just making an observation. The military made us this way, right? Constant vigilance. "Hope for the best, but plan for the worst" is a phrase that veterans often say to me. The problem with the word "but"[25] is that it negates or minimizes everything before the word..." hope for the best" is less important than "plan for the worst." Even when planning training or a mission, we consider the most dangerous course of action, and plan for it; that's why we have five claymores an AT-4 in the back of the vehicle. Just. In. Case.

We often carry that into our post-military lives, where it seeps into a lot of the different things that we do. Sometimes, we just get stuck in the negative, painful, or unpleasant. A friend recently sent a meme that pokes fun at it, but like many things (as DeNiro said in The Untouchables), "We laugh because it's funny...and we laugh because it's true."

Is that mental collapse inevitable, though? Not necessarily. I've written about how we can use concepts of positive psychology[26] during military transition, and this is along those same lines. One of the things about pessimism is that we forget that we have the ability to deliberately replace a negative outlook with optimism. That's what the title of this post means: for every mile of darkness, we need an equal distance of light in our lives to balance that out. Perhaps more, even; for every one mile of darkness, travel two miles of light. We can, and should, do this in order to live the post-military life we desire. To not continue to be in pain. No not continue to suffer.

Why Do We Focus on Negativity?

As I mention above, it's ingrained in us from the beginning of our time in the military. Medics are taught to search and treat the most catastrophic wound first. Infantry seeks out the enemy. Even in logistics, we anticipate potential problems and apply solutions even before the problem happens. All of this simply compounds our brain's inherent negativity bias[27], the tendency to focus more on negative things than positive things. That's something that has kept us alive, certainly, and absolutely when a veteran was in the military and deployed to a combat zone...but a tendency that is less beneficial when we are sitting in a barbecue with friends and family.

Why Should We Seek Positivity?

It is simply a fact that unpleasant emotions aren't always enjoyable. Sadness. Disgust. Fear. Even anger, which some (many) veterans enjoy, gets in the way of things we want to do. Sure, there are reasons for these emotions; no one is saying don't feel them. When we experience a loss, we should absolutely feel sad. The problem happens when that sadness does not dissipate, and turns into despair or depression. Or the fear turns to terror. More than that, though, studies show[28] that intense, prolonged negative emotions can cause physiological changes as well: high blood pressure. Hypertension. These emotions are associated with changes in our neural pathways, our cardiovascular and endocrine systems, and even our muscles. Prolonged negative emotion is not just psychological painful, it can be physically painful as well.

Does It Even Work?

If you believe it does, sure. If you think it's crap, then deliberately focusing on positive things to counterbalance negative things won't be effective for you. That's how the mindset works: as Henry Ford famously said, "if you think you can, or think you can't, you're right." The study linked above gives documented proof that it *does* work: positive emotion broadens our thinking and our behavior. They increase our mental flexibility, develop stronger coping techniques within us, and motivate engagement in enjoyable social activities. Most important is the fact that, as we start to feel positive emotion, we want to experience more of it, creating an upward spiral of wellness and resilience (Garland, Fredrickson, Kring, Johnson, Meyer, & Penn, 2010)[29].

How Do We Not Do It?

This is not just some "look on the bright side" Pollyanna optimism; that kind of thinking often minimizes the importance of the event that caused the painful emotion. Things are going wrong at work…" well, at least you have a job." Brené Brown[30] has a great short video that shows the difference between sympathy and empathy, and the dangers of trying to simply create a silver lining around a bad situation

So How Do We Do It?

Like many other things that we're not used to doing, we need to do it deliberately. We have to consciously search for, and experience, positive things to counteract negative things that are happening. The Army's Master Resilience Training teaches soldiers to Hunt the Good Stuff[31]: deliberately focusing on, and analyzing, positive things that have happened within a period of time. Again, it's not simply wiping out the negative; it's acknowledging the negative while also acknowledging the positive. It's a key part of the awareness and accountability log[32] that I developed: identifying three good things that happened during the day. Deliberately seeking out positive things or relationships we have in our life, and reflecting on them there.

The cycle between positive and negative can be repetitive: we think a negative thought, and then find a positive thought to counterbalance it. We then say, "yeah, but" and follow it up with a negative thought. Sort of like a roulette wheel...wherever the cycle stops, that's what you're going to feel. Stop on negative, experience unpleasant emotions; stop on positive, experience pleasant emotions. It's as easy...and as difficult...as that.

Consider these phrases as you go through your day:

For every moment of despair, search for a moment of hope
For every moment of sadness, search for a moment of joy
For every moment of terror, search for a moment of calm
For every moment of anger, search for a moment of peace
For every moment of pain, search for a moment of pleasure

Perhaps, in this way, we can start to allow the darkness to lift and the light to take over.

Bridging the Gap Between Beautiful Lies and Ugly Truth

"Life is short and truth works far and lives long: let us speak the truth." — *Arthur Schopenhauer*

If there's anything that I value, it's the truth. The painful, brutal, unvarnished truth. When faced with a choice between a beautiful lie and ugly truth, many choose the beautiful lie. I prefer the ugly truth. Sometimes it hurts. Sometimes it's an obstacle to getting done what we want done. It's never wrong, by it's very definition, but it's not always pleasant. The truth I'm talking about today is that there are aspects of military service that make it less great than it seems to be. Many veterans have an idealized version of their time in service, but the truth is much more accurate than our nostalgic memories.

"Why air our dirty laundry," some may say. "Us veterans have a hard enough time without people knowing what really happens in the military." I hear this often when it comes to veteran mental health; one veteran reached out to me after reading the article about the Pandora's Box of the Veteran Mind[33]. "It is very difficult to attain meaningful and profitable employment for those of us fighting the 'Broken Vet' stigma," the veteran said. "Telling the world we are a "Pandora's Box" does not help."

I understand. I really do. It's hard to walk a line between talking about the very real challenges that veterans face, in order to help them overcome them, and to not perpetuate the impression that all veterans are crazy killers. And not all articles are about the negative aspects of mental health; but by only presenting puppies and rainbows is as inauthentic as only posting doom and gloom.

Here are some truths about military service. They're from my own experience, or what I've seen from veterans I work with as a therapist. They're real. These things impact our mental health and wellness when we're in the service, and carry them into our post-military lives. And life is too short not to address the truth.

Truth: Racial Discrimination Exists

I have heard about some of the most overt and horrendous racism that anyone can endure. A Hispanic veteran being told that their job in the motor pool is to sweep and take out the trash "because that's all you people are good for." Another veteran who, while on deployment, talking about the dream of getting a ranch for him and his family, and is told…buy a supervisor…" you can't do that, you only get forty acres and a mule." This is not saying that all service members are racist, or even that most service members are racist; but this does exist in the military. It does happen, and it impacts our fellow service members.

Leaders, especially good ones, may not want to think this is happening. I know that it burns me up inside to hear it; even in a therapy session, admittedly, the leader in me reacts angrily to think that this is still happening in the military today. And when it does happen, and the service member reports it and nothing is done, then the negative impact of the discrimination is increased. Multiple studies, reviewed in this article in 2000[34] about the impact of racism on mental health and wellness, identified both physical and psychological distress as a result of racial discrimination.

Truth: Sexism, Sexual Harassment, and Assault Exist.

A colleague, Meaghan Mobbs, recently wrote about the stereotypes[35] that female veterans are expected to adhere to the military, and she hit the nail on the head. I once had a veteran tell me that she just got used to the fact that, when she went to make copies, she would be brushed up against or "accidentally" touched. "Why didn't you do something about it?" someone asked. She said, "If I did that every time it happened, I would be fighting all the time, every day. You reserve your indignation for the really blatant stuff." And, of course, many have heard about the Marines United scandal[36], referenced in Meaghan's article. A group of service members posted explicit photos of their female counterparts from all branches. This isn't the 1950s where the boss has the secretary make coffee and sit on their lap after making coffee. This is happening right now, today.

As a leader in the military that worked with females my entire career, I certainly did what I could to foster an environment where this was unacceptable. Even when I had a soldier come to me to report systematic harassment and unwanted advances, and we reported it, nothing happened. An investigation was conducted, it was implied that we were trying to "make the unit look bad," and ultimately both her and the perpetrator were moved to different units. Not surprisingly, research shows[37] that these types of incidents lead to greater incidents of depression, anxiety, and stress in those who experience them.

Truth: Toxic Leadership Impacts Veteran Mental Health

As horrendous as the sexual harassment and racial discrimination are, toxic leadership in general exists and has an impact on mental health and wellness. In this article from 2004[38], these are the characteristics of a toxic leader: 1. An apparent lack of concern for the well- being of subordinates. 2. A personality or interpersonal technique that negatively affects organizational climate. 3. A conviction by subordinates that the leader is motivated primarily by self-interest. For those of you who have served: sound familiar? I know it does for me. And what happens when someone is faced with an adverse situation that they can't avoid? Learned helplessness[39] starts to set in. Involuntary Defeat Strategy[40]. These, in turn, lead to depression, anxiety and stress.

These are the ugly truths. And there are certainly more of them out there. Again, are all veterans like this? Or even most? No, but chances are, if you are a female veteran or a racial minority, you have experienced something like this. And whenever I talk about toxic leadership, nearly every service member recognizes multiple leaders that negatively impacted them. These situations have long-lasting impacts on veteran mental health, and until they are addressed, will continue to do so.

Ambiguity and Lack of Feedback in Post-Military Life

One problem that many veterans face in their post-military life is how to manage ambiguity. I've experienced it myself, and seen many other veterans struggle with it. It's not that we always need the black and white answer, just that we're not used to *no* answer. "Give it to me straight" is the request. let me know what's going on so I can decide how I'm going to respond.

In my opinion, this is one of the central problems when it comes to employment searches. In the military, especially at the lower ranks, you were given immediate feedback about whether or not you "get the job." In the Army, through the promotion board system, you don't have to wait for a call weeks later (if at all) to see if you "aced the interview." You are told within hours of it happening, if not minutes. When job searches…not so much. "We've decided not to move forward with your application at this time" might be the most feedback that you're going to get.

This can an even greater challenge if there are other things involved, like a veteran working through some traumatic stress reaction. One study found that combat veterans who are diagnosed with PTSD are "significantly more adverse to ambiguity…but only when making choices associated with potentially negative outcomes" (see the full study here[41]). This makes sense, of course, if a veteran is actively looking to avoid negative outcomes, which is a central component of posttraumatic stress.

There are some factors to consider when it comes to managing ambiguity in post-military life:

Feedback is a Significant Part of Military Service

For service members, feedback was an integral part of the experience. Formally or informally, we were routinely told how they were doing in general, or specifically on certain tasks. You were either a "go" at this station or a "no-go" at this station. You were never a "neither" and then left to figure it out on your own. Go" at this station? Move out, troop, and get the next thing done. "No-go"? Remedial training, do it again until you got it right. Effective training in the military was a "tell-show-do" process. Tell me what you need me to do (classroom portion). Show me what you need me to do (practical exercise). Have me do it until I get it right (evaluation). Every military school I ever attended, or training event that I was a part of, had these components to a greater or lesser degree.

This doesn't happen in post military life. There might not even be much "telling," there is rarely any "showing" and the "do" is expected. This takes some getting used to, and is part of adapting from a culture of feedback and reinforcement to one of ambiguity.

Being Uncomfortable Not Getting Feedback Is Not the Same Thing as Needing Feedback

Many veterans I know get frustrated by the lack of feedback they get. I know, because I experience it myself. Just because it makes us uncomfortable, however, doesn't mean that we need it to survive. This is one of the "background programs" that we don't realize is there until it's gone. Sometimes, we don't even realize it then. It's something that we can adapt to, often because we will have to adapt to it. We're simply not going to get the task-oriented feedback in post-military life that is the same as we got while we were in.

Just because we want it, though, doesn't mean we need it. We don't always need to be told what to do; service members are adaptable enough to manage their expectations. If we expect feedback, and don't get it, then we get frustrated. If we manage our expectations, then we're less likely to get frustrated. We can do that by anticipating ambiguity.

Veterans Can Learn to Give Themselves Their Own Feedback

For me personally, and veterans I work with, getting any type of feedback is better than none at all. If it's a "yes," I know what to do next. If it's a "no," I know what to do next. When it comes to just waiting, however, then I don't have a third option. We can give ourselves that third option, however; in the absence of feedback, move forward with what we know. In the face of uncertainty, make a decision based on the most accurate data available. If the only data that you have is that you worked hard to prepare for something, did your homework, made every effort to be prepared…then rely on that knowledge and be satisfied with your effort.

Another way that a veteran can give themselves feedback is to find a trusted mentor. Someone who is invested in you and your personal and professional development. If it's employment, or job performance, or whatever, finding someone who is willing to tell you that the emperor ain't got no clothes is a way to know how you're doing. If it's a resume tweak or a technique that needs to be improved, then a mentor can help you make the adjustments. The mentor might come back and tell you that all looks well…then it's the other party's problem, not yours.

Understanding that feedback was a critical part of military service, and that it will not necessarily be so in post-military life, can reduce a lot of the frustration that we experience in life after the military.

Seeking Serendipity: Paying Attention to Beneficial Chance

Look for something, find something else, and realize that what you've found is more suited to your needs than what you thought you were looking for – Lawrence Block

There is an aspect of serendipity that is critical to satisfaction in life after the military. "Serendipity" is one of those words that we kind of think we know what it means, but not really. The quote above gives an accurate definition: serendipity is finding something you weren't looking for that has a positive outcome. It's paying attention to the world around you. Being aware of what you want rather than seeking what you think you need.

There is often a sense of dissatisfaction in post-military life. Sometimes, this happens when a former service member simply carries on with what they did in the military. Not considering other options; that's anti-serendipity. I work with many veterans who aren't satisfied with their post-military jobs; some, because they took the first thing that came along that paid the bills. Others, because they were just going along with what they were already knew. The cook becomes a chef. The logistics manager runs a warehouse, the intelligence analyst becomes a government contractor, the leader becomes a corporate leadership trainer. And something is *missing*.

What's missing is the disconnect between what we think we want to do and what we really want to do. Serendipity can be the link between the two. We only have to pay attention to what the universe is telling us.

We can't force serendipity to happen. What we can do, though, is stop forcing service members down a pre-laid path. Google developed a search tool in which a veteran can put in the job they did in the military, and they would get non-military job recommendations. My colleague David Lee wrote an article about how they got it wrong[42]: funneling service members down the same path they were on in the military.

In order for serendipity to work, we have to be aware of our own desires and inclination, which takes self-awareness. We have to develop that within ourselves. And, like many other things, research has shown how to make serendipity happen. An analysis of serendipitous events[43] show that there are three things necessary for it to happen: a chance encounter, a prepared mind, and an act of noticing.

Serendipity Has to Happen by Chance

Serendipity is a chance encounter with something we weren't looking for that turns out to be great. When I punch in my former occupational specialty in the Google MOS translator, I get a series of jobs that tells me I would be a good truck driver. I knew that's not what I wanted to do, though. People often ask me how I became a mental health counselor; my answer to them is serendipity. I happened to be attending a post-deployment reintegration seminar, in which the person giving the class on relationships said, "By the way…anyone who is interested in psychology, consider a career in the mental health field after you get out of the military. There are not enough combat veterans in the industry."

That statement is what led me down the path I'm currently on. It happened by chance; that statement wasn't part of her class. I could have been in another group, or she could have not been a part of the whole seminar. I was one of a hundred or so folks in that room; to my knowledge, I was the only one who has become a therapist.

Even though we can't force serendipity to happen, we have to be prepared for it. I did happen to be interested in psychology. Taking care of troops is what I did in the military; this is just an extension of that. I knew that I didn't want to continue to do what I was doing, but wasn't sure what it was that I wanted to do.

Serendipity Requires a Prepared Mind

When it comes to taking advantage of beneficial chance, we have to be open and ready for it. Our minds have to be in receive mode, listing to what we want in our hearts and what the world is presenting to us. This is one of the biggest problems that I see with taking advantage of serendipity; we have our head down, blinders on, focused only on the path in front of us and not even listening for other opportunities.

A prepared mind has two elements, according to the article linked above. First, there must be a prior concern. There has to be a need to be filled; a dissatisfaction with the way things are, or a problem to be solved. Second, there has to be previous experience or expertise. Not necessarily formal training, but familiarity with the subject. Both of these things require awareness; if we are not aware of the dissatisfaction, and we don't have previous experience, then we're not able to see opportunities that may be there. We may not be able to cause chance to happen, but we can improve our odds by preparing our mind.

Serendipity Requires An Act of Noticing

After having a prepared mind…knowing that I didn't want to continue in logistics, and having the experience of leading and counseling soldiers…we have to notice the world around us. We can be as prepared as we want, but if we ignore stuff, then we're not going to take advantage of serendipity. Noticing requires us to pay attention; again, if we're walking the path blindly, the opportunities won't even register in our minds.

Another problem to avoid is when we talk ourselves out of something. If I had heard that statement and was stuck in the doubt trap, I wouldn't have taken advantage of it. *"Me? A therapist? There's no way I could do that."* We can get in our own way, and not get the beneficial effects of serendipity, if we ignore the signs that something else might be there for us.

How has a random chance encounter benefitted you or your family in your post-military life? I'd love to hear them. Drop me a line or comment below; more examples of beneficial chance can help other veterans and military family members prepare their mind, notice the opportunities, and take advantage of the serendipitous moments in our life.

Thankfulness for Transition and Change

There's a lot of talk about the "transition process" when it comes to service members and veterans. Of course it's significant; leaving the military is shifting from one part of our life to another. Whether we serve three years or twenty-three, we move on from one way of life to adapt to a different one. For some it's easy; for others, it's a daunting challenge, like crossing over the "bridge" in the picture. Why, though? What makes things so challenging for service members as they move from the service to post-military life?

Personally, I'm thankful for the transitions I've experienced. If we're not growing, we're not living.

Think about it. We're constantly in transition. From being a toddler to childhood, from adolescence to adulthood. Moving from our hometown, or moving back to our hometown. Moving out of...and into...our parent's basement. Consider our time in the military...there was near constant change. When I arrived at Fort Carson in 2006, I was one of the first soldiers in the company. We had five pieces of equipment. Ten months later, in October, we deployed with a company strength of over 130. The year before, none of us knew each other; one year later, we were deploying to combat. If that's not a significant transition, then I don't know what is.

So what is it about the transition out of the military that is so challenging? How do we approach it so that it's easier on us? Here are some thoughts.

Transitions Are Natural

We're not going to avoid change. The plains meet the mountains, the sand meets the ocean. There is a shifting transition between the seasons, and there are moments when we realize that we are no longer children. Sometimes, the transitions are clear; a marriage, leaving for the military. Many combat veterans remember the transition from "I may get shot at and have to shoot people" to "they're shooting at ME." At other times, the transition is more general. We wake up one day and we're fifteen years older. The step is a little bit slower, the grey hair is popping out just a little bit more.

Change can come smoothly, like slipping into a pool, or it can come with turbulence, like the rapids on a river[44]. Navigating the smooth transitions doesn't take a lot of skill, but navigating the choppy waters sure does. We have to develop the skills ahead of time...be prepared for the change before it happens...or develop the skills in the middle of the change. If we don't we risk being overwhelmed.

Our Mindset Can Make Transition Challenging

This isn't "when life gives you lemons" type of thinking, some blind optimism in the face of obvious difficulty. If we don't acknowledge change, however, then it's going to be difficult. It's almost as if we tell ourselves, "Okay, here's a transition; we can do this the easy way, or we can do this the hard way." Consider two paths to the same goal. We're standing at one point, and have to get to the other. One path leads across an open field, an easy walk to get to where we need to be. The other path leads down and around, the long way, and it's filled with obstacles.

Which do you think veterans choose? If they're like many of the ones I see as a mental health counselor...or like me...they'll take the long obstacle-filled route. Because anything worth doing is worth doing difficultly. That mindset makes the transition more difficult that it needs to be.

Transitions Can Be Managed

It can be done. There are veterans that I served with, at the same time and on the same battlefield, that struggled with their transition more than I did with mine. I haven't been unemployed since I retired; my family remains intact, and actually stronger than ever. I sometimes think to myself, what was different for me than it was for them? My faith, the fact that I was older during my deployments, the fact that I have military retirement pay to fall back on? I know folks who were older than me, had as much faith as I do, who have stable incomes. They still struggle. How have I been able to manage my transition, while others have struggled? Again, I think it comes down to mindset.

I recently had the opportunity to attend a leadership development program with two double amputees. I think to myself; what makes them different than another veteran who, like Lieutenant Dan, is bitter and resentful? I can only come up with the idea of mindset. We control how we see the world[45]. Even in the movie, Lieutenant Dan comes to a measure of peace after a period of time; but why does it have to take so long?

Support Can Make Challenging Transition Easier

Transitions are going to happen. Change is going to occur. It's about mindset, and the difficulty can be managed. It often can't be managed alone; sometimes, changes are more difficult and we need to call for backup. When my father passed away, one of the first things I did was call my therapist to set up a meeting. I reached out before things got too much. And that's the key; if we find transition to be overwhelming, we don't have to do it alone. There's nothing wrong, and everything right, with reaching out before a crisis happens.

Avoiding Avoidance Leads to Success in Post-Military Life

"Success doesn't mean happy every day." – Omar Andrews

I don't wanna, and I ain't gonna. How often do we avoid unpleasant experiences so we don't feel uncomfortable…but end up not resolving the issue, and turn pain into suffering? Avoidance is a natural reaction to unpleasant situations; we've been doing it since we were kids and spit out the mushed beets our mothers were trying to feed us. When it comes to veteran mental health, however, avoidance can do more harm than good.

As I've mentioned before, I have found a concept called Dialectical Behavior Therapy to be effective with the veterans I work with in clinical practice. Of the many skills that I have found effective, the one that applies to the concept of avoidance is the practice of opposite action; acting in a way that is in opposition to the way that we feel. This is called, rightly enough, "avoid avoiding." Rather than back away from conflict, address it head on; stop avoiding it. This doesn't mean flip tables because you want to, but if we have a tendency to back away from something, then that means that there's something important going on. We should, instead, acknowledge the problem and resolve it.

Avoidance is Not Something Veterans Did in the Military

One of the things about avoidance; it's another thing that we typically didn't do in the military. Why should it crop up in post-military life? It didn't often happen in the military, because it generally wasn't allowed. We didn't get to avoid the unpleasant feeling of standing watch in the rain, or running the obstacle course for the tenth time. In many ways, that helped us get over the hesitation that comes with avoidance; to get us used to doing things that we would normally shy away from. Heck, that's what the greatest pavlovian conditioning experiment in the world…Airborne School…does for you. You throw yourself to the ground, over and over again, from successively greater heights, for one single purpose: when a little green light goes on, you jump out of an airplane.

In highly dangerous situations, it's even more pronounced. You run towards the sound of the bullets, not away from them. Firefighters run up the stairs in the World Trade Center when others are running down. Through training, and determination, we avoided avoiding danger and conflict in the military; why do we engage in avoidance after the military?

Avoidance Works in the Short Term

The challenge with avoidance is that it works. It makes life easier, just in that space of time. When we decide not to do something unpleasant that we know we need to do, we get a temporary feeling of relief. We know we have to fix up that spare room, or clean up our desk, or confront that coworker that we know is doing something wrong. We just don't want to experience the unpleasantness. And so that can be relieving...but it doesn't solve the problem. The spare room doesn't get fixed, the desk is still chaotic, and our coworker is still doing what they shouldn't be doing. In the long run, avoidance only creates more problems. It's like emotional procrastination; why do today what you can put off until tomorrow? Because the problem is still going to be there, whether it's today or tomorrow.

Avoidance Complicates the Long Term

Like many things, however, the initial benefit covers up the more significant long term problem, and it impacts our emotion and behavior. We know we need to take care of that unpaid bill, but we don't even open it...just leave it lying there. We avoid dealing with it, and every time we look at it, our anxiety increases. The longer we let it go, the worse it gets. The urge to avoid keeps us from addressing something unpleasant, but it delays the consequence of what we're trying avoid. We know that we need to get help with what's going on in our head; we can't sleep, we're drinking too much, arguing with our partner. Even after we realize that there's something not right, we stubbornly avoid reaching out for help. After all, reaching out for help is weakness, right? And the relief that comes from the avoidance of getting help is not more beneficial than addressing the problem head on.

Avoiding Avoidance Leads To Success

Sooner or later, the problems in our lives have to be dealt with. The shut door keeps us from moving forward; and moving forward is success. The quote at the beginning of this article was shared with me and a group of my colleagues by a former Marine who was finishing college and starting to work in legislative advocacy. Success doesn't mean happy every day...it means sometimes doing things we don't want to do. It means avoiding the urge to avoid those unpleasant experiences that, while unpleasant, will solve problems and get us on to the next task. The insidiousness of avoidance is that it is failure wrapped in the illusion of success; true success will not be gained until we start to avoid avoiding.

Easy? Of course not. Nothing worth doing ever is, every service member knows that. The safe and easy route is usually deceptively dangerous. We can take that too far[46], of course, and make life more challenging that it needs to be, but we don't need to avoid the challenging days entirely. Avoid avoiding, and find success.

Veterans and the Zone of Emotional Effectiveness

Is there such a thing as just enough of a "bad" thing? Or too much of a "good" thing? Does that make the bad thing good, and the good thing bad? I try not to label things as good or bad, because that's a judgement call. Here, I try even harder not to label; I'm talking about emotions. In my opinion, there are no such thing as "good" emotions and "bad" emotions. There is a zone of effectiveness when it comes to our emotions. Too much of them is not good, and too little of them is not good, but quite often, just the right amount of emotion is beneficial.

Emotions are necessary, regardless of what your drill sergeant told you. As a matter of fact, the primary emotion that he or she displayed…anger…was extremely necessary for that time and that place. Without anger, without intensity, boot camp would not have been the test of your limits that it actually was. I've talked before[47] about how "bad" emotions…anger, fear…were effective and necessary in combat. We talk about people who are "emotionally dead" and "soulless." That's the part where there are too little of them. Then you have the overreaction, when our emotional responses are much too intense when it comes to the situation. When someone cuts us off while driving, frustration is appropriate. Maniacal rage is not.

In the Dialectical Behavior Therapy Skills Training Manual[48], Marsha Linehan describes three functions of emotions: we use them to communicate to and influence others, to organize and motivate action, and use them for self-validation.

Emotions Communicate to and Influence Others

Back to the drill sergeant: the expression of anger (although it likely really wasn't there, it was kind of an act) was to communicate to you the urgency of what needed to occur. And to influence you, of course. I'm not talking about using anger as a weapon to bully someone, but an angry response, or hurt response to something communicates that we don't like whatever just happened. Emotions are an important part of our interpersonal interaction. Anger can indicate to someone that something we believe is true has been or is about to be violated. Fear can indicate that there is danger in our environment, and expressing that fear alerts others to that danger.

By suppressing the expressing of our emotions…as service members are all but explicitly trained to do…we reduce the effectiveness of our communication to others. People don't know that they've crossed a boundary if we don't show some form of displeasure. People don't know that there is danger if we don't express some form of uneasiness. Similarly, people may misunderstand something if we show *too much* emotion, disproportionate from the environment. Showing the right amount of emotion for the right situation can help us communicate to others what we feel and what we want.

Emotions Organize and Motivate to Action

Emotions are also important in motivating us to do something: this is where we approach the zone of emotional effectiveness. Anxiety, many would say, is bad. It gets in the way, it paralyzes us. But that's if there's *too* much anxiety; a couple of research articles show that just the right amount of anxiety could be a good thing for us. One shows that an appropriate amount of anxiety is important in assessing and avoiding risky behavior[49], and another shows that anxiety and worrying can be beneficial in getting us ready for an important task[50]. The same could be said for anger; if someone cheats us, or insults someone we care about, anger is an appropriate response to that. Sadness wouldn't be, and happiness certainly wouldn't be. An effective amount of anger would allow us to express our displeasure, to stand up for what we believe; to let someone know that a line has been crossed. Just as an excessive amount of anxiety can be detrimental, as described in the second study above, and excessive amount of anger, or fear, can also be detrimental.

Similarly, I equate fear to jumping out of airplanes. In full disclosure, I was terrified of heights as a kid. My family knew it, it was sort of "my thing." They did all they could to "cure" me of it, which, of course, made it worse. So what did I do when I grew up? Decided to jump out of airplane. Repeatedly. And not just jump out of them, but become a Jumpmaster, whose job it is to stand in the door of the plane and stick my head out. Talk about fear…I knew it, I used it. I had an old Jumpmaster mentor tell us one time, "the minute you stop being afraid of jumping out of an airplane is dangerous. It means that you no longer respect the danger." TOO much fear, of course, could have paralyzed me, and too little fear would have made me complacent. Again, just the right amount is just the right amount.

Emotions Can Be Self-Validating

When it comes to self-validation, we often don't understand our emotions if we're not aware of them. Sort of like Spidey-Sense, emotions are the combination of physical reactions and psychological reactions to a certain event. If we feel anxiety about something, it signals to us that what we are about to do is important, and we need to pay attention to it; then, once we do it, the anxiety goes away. That is a signal to us that what we just did was important to us and we need to pay attention to it. Anger, similarly, signals to us that something we care about has been or will be violated. Our safety, or the safety of something we care about. When we respond to that threat...appropriately...then we resolve that situation, and anger goes away. Again, validation that our emotion was appropriate for the situation. After time, interpreting and implementing your emotions effectively becomes something that you rely on to make decisions. The "gut instinct" is critical for many, and is rooted in emotional awareness.

Emotions are necessary, or we wouldn't have them. Just like ice cream, or that unstable ex of yours, too much of them can be a very, very bad thing. Understanding how to find the appropriate emotional response to the event in your environment is the key to finding balance and operating in the zone of emotional effectiveness.

The Myth of Limited Resources of Support for Veterans

This is something I hear often from veterans that I work with: "I don't get help, because others are worse off than me." "What do I have to complain about? I made it back, didn't I?" "I don't want to take up your time, there are other veterans who need to talk to you more than me." While the "I'm not worthy" thing may work in Wayne's World, it does nothing but cause a veteran to continue to suffer needlessly. As my colleague Erin Fowler[51], says, "I'm the one who gets to decide when someone's wasting my time, and you're not it!"

It's almost as if veterans think that any resource that is available to them…mental health counseling, disability benefits, education benefits, housing or employment support…is limited in both scope and scale. It's like we think that these resources are like the food in our fridge: there's a limited amount, and when it's gone, then it's gone. "Better that I should go hungry," the sacrificial veteran says, "Than one of my brothers or sisters miss out on what they need."

Instead, these resources are literally like an All-You-Can-Eat Buffet. Not one of the sketchy ones that always seem to crop up outside the back gate of post, but one of the really good ones. The kind of buffet that has endless supplies of prime rib and lobster and a taco bar. You can think that you're saving a place for someone else, or that you don't deserve to eat at the buffet, or any number of things, but that doesn't make them true. The only person you are really hurting is yourself.

These Resources are Your Right, Earned by Your Service

By your service in the military, you have lifetime access to the full range and amount of benefits you've earned. Sure, different levels and types of service either allow access or restrict access to different portions of the buffet, but many veterans disqualify themselves from these benefits before even exploring whether they're an option or not. "I don't deserve disability benefits, look at me…I have all my fingers and toes." While, in the meantime, it takes you fifteen minutes to get out of bed in the morning because the shooting pain in your back or knees. Or your emotions are so varied that you don't know how you'll be feeling from one moment to the next. This is not "perpetuating the stereotype of the broken warrior," this is acknowledging that military service is a significantly dangerous and demanding occupation, both physically and psychologically. Your right to your benefits is like your right to vote: if you don't use it, the right doesn't go to someone else. It belongs to you, and you alone.

These Resources are Not Limited

Sure, there are rules and policies that restrict which benefits you can use, but usually by the time that you use them, they're as much as you need. Sure, I used my GI Bill to get a second Master's Degree; it would be hard for me to make an argument that I need to use Vocational Rehabilitation money to get a third one. But that's not me taking away from someone else…that would be me trying to take more than I need (see below). But just like the great buffet, the resources don't run out…they're there. When you walk up and fill your plate, there is some there, and when you go back, there is still some there. Even if you are standing in front of the buffet saying, "No, that's okay, I'll let you go first," and your battle buddy goes up to fill their plate, the resources are *still there*. There are benefits stacked on top of benefits that you earned by the right of your service. If you don't use them, it's not like they're going to someone else…they're endless, with some exceptions, but those are extremely rare.

Sometimes, Others Take More Than They Need

This happens. You know it does. There is always that one person at the buffet, with a plate in each hand and a third balancing on their forearm. They are literally taking everything that they can eat, and more, with apparent lack of regard for anyone else trying to get the same stuff. And do you know what? Even after they walk away from the buffet, satisfied down to the rolls stuffed in their pocket, there are still benefits available for you. Are there attempts to defraud the system? Yes, absolutely, and they're horrible. There are certainly some who have not earned benefits that are receiving them. But do you know what else? That's the government's problem, not yours.

I was First Sergeant of a company in 2010-2011 where we had the responsibility to help wounded soldiers transition out of the military. A senior leader in my chain of command, during a private conversation, said, "Look at all of these Soldiers. It's going to be our tax dollars that are supporting them into the future." My response to them: if every one of these Soldiers immediately recovered and returned to duty, would we stop paying taxes? No! I will acknowledge that maybe 20% of the soldiers in my company had simply got tired of the Army and were using this as an easy way out, but 80% of them were those who had served honorably and had legitimate wounds. The problem was, and still is, that the 80% is treated like the 20%; for me and my commander, we had to treat the 20% as if they were the 80% so that the majority of them got what they deserved.

Not Accessing Resources Hurts You, and No One Else Benefits

So there you are, standing in the middle of the buffet. Starving at the feast[52]. People are handing you food, begging you to take it, practically filling your pockets with it, and you say, "Nope. No thank you. That should go to someone who needs it more than me." Guess what? *It won't.* It will just sit there, waiting for you to access it. Others…like veterans with undeserved bad paper…would be thankful for even a tenth of the resources that are available. By rejecting the resources that are yours by right, you're perpetuating the struggles that you might have. Is the suck so good that you want to embrace it all the time, even when you don't have to?

Don't wait. Don't think that others deserve it more than you. Sure they deserve it…and so do you. Reach out and take advantage of the things you've earned, and live the post-military life of peace that you've always wanted.

Ten Reasons for Veterans (Not) To Avoid Mental Health Counseling

No way, pal. You won't catch me going to the therapist. I'm going to avoid that crap so hard that it hurts. I'm so focused on holding it together that I don't have time to yap at someone who's only going to judge me anyway. No thanks, I'll handle it in my own way.

How many times have you heard that? How many times have you *said* that? The maximum effective range of an excuse is zero meters, as one of my leaders used to say. These reasons for avoiding mental health counseling are just that: excuses. They are effective in keeping us stuck in a crap way of thinking and living, but they are certainly not effective in helping us live the peaceful post-military life we want. Here are ten reasons for avoiding mental health counseling that I've heard.

It Doesn't Help

This goes in the category of "it's going to stay this way forever." One of the main reasons I hear from veterans is that they don't understand how talking to a mental health professional can actually help them come to terms with what they experienced. The frustrating thing for us mental health professionals: we know it works. We see it every day. We know veterans whose lives have been changed, very much for the better, once they started talking to a mental health counselor and addressing their thoughts, behaviors, and emotions. It does work, you just have to try.

I Can't Talk To Them if They Haven't Been There

This is a big one. "If they're not a veteran," or even some more specific variation of that…a combat veteran, a combat arms veteran, a veteran from the same era/service/occupational specialty as me…then "they don't get it." As a combat veteran, sure, there is a bit more of a fast track to trust with my clients. That doesn't mean that you have to have served to be able to help someone who has served. First, there's simply not enough combat veterans in the clinical mental health field to meet the need. Second, a licensed mental health professional has a level of training and expertise to address mental health concerns, and if they're working with veterans, then hopefully they've done some work around understanding the military and veteran culture.

If you're unsure, then ask: do you work with a lot of veterans? If they say, "I'm just starting out, and I don't really know that much about veterans," then maybe you want to find someone else. Sort of like when you take your car to the shop: "do you know much about Jeeps?" If they say, "I'm just starting out, I don't really work on them that much" you can rest assured that I'm not leaving my Jeep there. That does not mean that's the case with *all* mechanics, though.

I'll Lose My _____ (guns, security clearance, etc)

"If I go to therapy, I'll lose my security clearance" or "they'll take my guns away from me." Not true, but you have to believe it's not true before you try it out. Going to see a therapist doesn't mean you won't be allowed to own a gun. I see veterans as a therapist all the time, and they are able to hunt, go to the range, shoot skeet, all the rest of it. And security clearances? In 2016, the Director of National Intelligence implemented a change[53] to the infamous "Question 21" on the SF 86. The changes shift the focus from whether or not the applicant sought treatment, to whether or not a diagnosed mental health condition impacts their judgement, reliability, or trustworthiness. Even before that, though, seeking mental health counseling wasn't a disqualifying factor. After each of my first two deployments, my wife and I went to marriage counseling. That was in 2008 and 2010…and I applied for and was granted a Top Secret clearance in 2012. It didn't hold me back, and it shouldn't hold you back.

What's In The Past is In The Past

This is the "locked duffel bag" argument. What's in the past is in the past, it stays in the past, and will always remain in the past. Until it doesn't. That's the sneaky part about unresolved trauma or mental knots: they crop up in the present like yesterday's spicy food. We may have all of the stuff that bothered us locked away in a dark room of our mind, but that doesn't mean it's going to stay there. Ignoring it only makes it build up, and ultimately potentially overwhelm us. Or come out at weird times, when we don't want it to. Ignoring the warning light on the car doesn't make the problem go away, and ignoring the warning signs of unresolved pain doesn't make the problem go away either.

All They're Going to Do is Medicate Me

This is another common misconception. I've been doing this for years, and I've not medicated anyone. I'm not allowed to as a Masters level mental health counselor; that's Psychiatrists and Psychiatric Nurse Practitioners. Could medications help? Sure, maybe, depending on what you're dealing with. Are they always the answer? Not really, especially when it comes to aspects of veteran mental health that have no medication interventions, such as a lack of purpose and meaning, moral injury, needs fulfillment, or relationship challenges. There is no one-size-fits-all approach when it comes to mental health. Assuming that you don't like the one thing you think they're going to do before you see if they're going to do it is rejecting the answer before you even asked the question. Check it out and see.

I'm Doing Just Fine Without It

Are you? Really? You might be, but then again, you might be drinking a bit too much. Or yelling too much. Or getting too angry at those around you over small things, like perceived disrespect. There was a need to leave the boots at the door when we were in the military, and there is certainly a need to leave the boots at the door[54] when we leave the military. If you find your sleep wrecked by nightmares, or your days filled with anxiety or depression, are you really doing just fine without it? You might not be telling yourself the whole truth.

My Buddies Are Doing Just Fine Without It

And this one goes along with that one…are they really? Being vulnerable is not something that veterans are comfortable with. We want to show ourselves as capable, reliable. If we don't show what's going on behind the curtain to our buddies, what makes us think that they're going to show their reality to us? It's even worse with social media. We see our friends and their happy lives, eating great food, and going to interesting places…but it's been said that social media is like comparing our lives to someone else's highlight reel.

Going To Therapy is Admitting Weakness

This is a huge one. Probably should be up there near number one, if we were doing this in any kind of order. "If I go to a therapist, then it means I'm _____" and fill in the blank. Weak. A loser. Someone who can't hack it. Undependable. Worthless. How negative are each of those statements? How would we feel if we heard our kid say that about themselves? A study on the perceptions of stigma in veterans[55] asked a group of veterans to rate how they would feel about a particular statement, compared to how they would feel about their fellow veteran regarding the same statement. 44% of veterans thought that they would be perceived as weak for going to a mental health counselor; only 12% of veterans said that they would consider another veteran weak for doing the same thing. That's a HUGE disconnect…and one that keeps us stuck in the suck.

I've Tried Asking For Help Before

This is a challenging one. Yes, it happens. You've gone to the VA, like they said you should, and you can't get in to see someone for five weeks. Or you get sent to someone in the community, and they have no clue how to handle what's going on with you. I've heard it all: therapists who start crying when they hear you explain what you went through, or argue about your politics. I once had someone once tell me that the therapist they saw would sit on the floor, legs crossed, with their eyes closed during the session. I get it! But I'm not them, they're not me, and that's not even someone who represents the mental health profession. We don't stop at one mechanic when the car needs to be fixed, but we will absolutely stop at the first sign of someone who can't help with mental stuff. As Air Force veteran Rhi Guzelian said on a recent podcast[56], a mental health professional relationship is as much personal as it is professional. Find someone you get along with, and avoid those that you don't. It's as simple as that.

There's Nothing Wrong With Me

Similar to the "I'm doing just fine" argument, but different as well. I know that I'm not doing fine, I know I'm struggling, but what do you expect? Have you seen what I've gone through? There's nothing wrong with me that a little _____ can't fix. Fill in the blank again: booze, isolation, gym time, angry rant. In the meantime, after we keep applying these band-aids, the cracks keep getting bigger and bigger. "This too shall pass" may work in some cases, but definitely not the majority.

So there you have it; some of the excuses that I've heard for veterans avoiding mental health counseling. Any of them sound familiar? Any that I missed? Feel free to reach out on social media or reply in the comments below. Love to hear anything I forgot.

The Healing Power of a Stranger's Ear

This may seem odd, but at least three times a week, strangers tell me their secrets. When I meet a new veteran or military spouse in my office for the first time, I know very little about them. They know very little about me. Okay, in this digital age, they likely did their homework, and googled me, or asked someone about me. And I likely got some preliminary information about them. I'm certain about some things. Their connection to the military is a given, considering the work that I do. There is a desire to get help with whatever their problem is. Outside of those very limited circumstances, there is little that we know about each other, though. And, yet, they talk.

Maybe it's because I'm a therapist, and that's what you do with mental health professionals: talk. Maybe it's because my office looks like the picture above, because that's what it does look like: maps on the walls, coins on display. And, maybe, it's because when a veteran finally finds someone who is willing to listen to their story, they're ready to tell it.

There is something very liberating in unburdening ourselves of our thoughts. I've heard variations on a theme: "it feels like a weight has been lifted." "It feels better getting it out." Why to strangers, though? Why to a therapist, or a faith leader, or a friend? Why does that sometimes feel better than sharing it with those that are closest to us? It's not that our families don't care for us, because they really do. It might have a lot to do with the fact that they care for us very much, and the independent observation of an individual with less emotional connection to us is more objective than the observations of those closest to us.

Concept of Expressed Emotion

The psychological concept of Expressed Emotion is defined as a "measure of the family environment that is based on how the relatives of a psychiatric patient spontaneously talk about the patient." This 1998 article describes that the family can express hostility, criticism, and emotional over-involvement with their loved one who is experiencing psychological symptoms. This can even be more challenging when the family sees a veteran as one way before deployments or a traumatic event, and another way after.

Expressed Emotion and Veteran Mental Health

Some examples of how criticism can be expressed regarding mental health can be heard in this episode of the podcast Invisibilia. Criticism can take on the form of a parent complaining about the fact that their schizophrenic son does nothing but drink coffee and smoke cigarettes instead of eating breakfast. Or hostility towards them, describing the family member as selfish or uncaring. Emotional over-involvement is demonstrated by the family member saying things like, "I will do ANYTHING to help them get over this" and "It breaks my heart to see them suffering like this."

How does this apply to a veteran struggling with symptoms of posttraumatic stress, depression, or substance abuse related to their military service? I've heard variations on these themes, too. "Why can't you just go back to the way you were" is a form of expressed criticism. Blaming the problems in the family on the veteran's behavior is a form of hostility. And the phrase "I would do anything to help my veteran overcome this, I would sacrifice EVERYTHING" is emotional over-involvement. I've even heard this from people in the community who want to help veterans…that's where my often repeated description of veterans as "broken-winged birds and three-legged dogs" comes from. Someone actually said that to me once, that they love helping veterans, because they feel such pity for them. Wrong answer.

How Expressed Emotion Can Block Recovery

Treatment is often not supported in families or situations where there is high expressed emotion. In other words, when hostility, criticism, and emotional over-involvement are present, the less likely treatment is going to be effective. There has to be a balance, obviously; pointing out things that are not going well is not the same thing as criticism. Expressing ourselves in an emotionally balanced way is not showing hostility. And showing concern for a veteran is not the same thing as emotional over-involvement. Like in many things, too much of this is where things start to go wrong.

A veteran will often become defensive when faced with hostility, self-critical when confronted with criticism, and withdraw or resist emotional over-involvement. When we have an emotional bond with someone, our ability to remain objective is significantly in question. I, as a mental health professional, know that; I can't be my wife's therapist. Or my children's. Or my father's therapist, even though he was a Vietnam veteran. A colleague of mine, a competent psychologist who works for the VA, is also the mother of an OIF veteran. She often says, "when I talk to him, my psychologist hat isn't even in the room. My mom hat is on all day long." When we are emotionally close to someone, there is a greater chance that we might engage in one of the three expressed emotion domains.

How Strangers (Mental Health Professionals) Counteract Expressed Emotion

So how does the stranger play into all of this? Studies have shown that lower expressed emotion leads to greater chances of recovery. If a veteran is working with a mental health professional, they are hopefully not subject to criticism. If you're getting hostility from your therapist, then it's probably time to look for a new therapist. And the objectivity of less emotional involvement is precisely the benefit of a stranger's ear. Mental health professionals work at developing an accepting, non-judgmental attitude regarding the clients they're working with. I am not connected to all aspects of my client's lives, day in and day out. A stranger has the objectivity to be able to separate themselves from the situation, and offer observations that might be more beneficial than someone who is so emotionally close.

And sometimes, that objective observation is just the right thing to set us on the path of awareness and recovery.

The Misconception of the All-Knowing Therapist

There is a mysteriousness when it comes to therapy and mental health. Some might think that it's snake oil, some made up mumbo-jumbo that has people talking about their feelings and whining about how their papa didn't hug them enough. Then there's the idea that you're just going to lie on a couch while some dude with a goatee and a jacket with patches on the elbows of his jacket asks you about your mother.

All of these can contribute to the avoidance of therapy and the idea that we can handle things on our own. There is another misconception about psychotherapy that I've experienced, however: the idea that the therapist has all the answers. I have sometimes had veterans come into the office, sit down, and say, "Okay, I'm ready for you to therapatize me. Here's my brain, tell me what you see."

Therapy doesn't work that way. Sure, I have the clinical training in diagnosis and treatment and understanding about mental health, the mind, and the brain. I also have the benefit of seeing a lot of veterans throughout the week, and see the common experiences among all of them. A veteran described me once as an "experience nexus," in which I accumulate observations from many veterans and apply them to the particular veteran or military spouse in front of me.

There are some misconceptions that some veterans have, that therapy is something that is done *to* them rather than something that is done *with* them. Perhaps some of these points can bring some clarification.

Therapy is All About The Client

It's your brain, you have to do the work. I'm not an expert at much of anything, but I'm certainly not an expert on you or any other veteran sitting in my office. I do have to say, this is coming from my own personal theoretical orientation; I have a client-centered focus, developed by Carl Rogers[57] in the 1940s. In Rogers' paradigm, we move away from the idea of "practitioner as the expert" towards an understanding that the client is the focus, not the knowledge that the therapist has. A veteran that comes to see me is not "sick" or "broken" and needs me to heal or mend them; instead, sometimes we don't fire on all cylinders and could use some support in getting back on track.

Therapy is Not Mindreading

There are a lot of assumptions that we have when it comes to communication in general. It bugs my wife to no end when I finish thought a conversation in my head that I was talking about out loud. It's like I finish my sentences in my mind and then go on to the next thing, my mind moving too fast for my mouth. Other times, people start a conversation in mid-thought, starting a conversation in the middle of a thought as if the other person knew what they were talking about. There is a misconception that therapy is like that: the therapist can see into my head and tell me what's wrong there.

Don't get me wrong, anyone who observes human behavior can learn a lot just from watching. How we hold ourselves, the unconscious movements of our eyes or mouth. Just observing words or mannerisms can seem like mind reading, but it's not; it's just observation. We can learn to be as aware of these things in ourselves as we are aware of them in other people, and demystify the process.

Therapy Isn't Dispensing Answers like Medicine

As I mentioned above, therapy is something that is done *with* the veteran, not *to* the veteran. I often say to veterans, I'm not an answer machine...we have Google for that. Instead, I'm more of a mirror, I reflect back to the veteran what I see, what I observe, and they can tell me if it's accurate or not. Sometimes that alone is enough to help shed some light on a sticky situation, getting a different perspective[58] on a problem. What it isn't, however, is listening to someone for a short period of time and writing out on a piece of paper, "here's the problem. Here's how you fix it." It's a very real and honest conversation with someone you trust, and someone who has the clinical training and experience to help you understand more about yourself.

Therapy Isn't a Spectator Sport

This is something that often surprises me, that some veterans feel as though coming into the therapists office is like going into your doctor's office. You walk in, show them your foot or elbow or whatever, they run a few tests or take some pictures and say, "There's your problem." Instead, it's a two-way street; a knee is a knee is a knee, but the variability in the human mind is so great that there is simply no way to engage in therapy as a passive observer. There's not a diagnostic machine that we can hook up to our brain, like in our cars, that spits out a code and says, "this is what's wrong."

Instead, much of the benefit from therapy comes *outside* of the counseling session. When someone reflects on the things that were talked about, uses some of the coping techniques that were discussed. Mental health counseling is not a passive pastime where we just sit and have a conversation; we have to apply what we've learned to the real life situations outside of the room. Only then do we begin to take control of ourselves, and achieve what we know we're capable of.

We Can't Change What We Don't Talk About

What you talk about you can control, influence and change – David Osborn

As I was listening to an episode of the Follow Your Different podcast[59] hosted by Christopher Lochhead, I was struck by this statement by his guest. In that context, entrepreneur David Osborn was talking about money. He was referring to the fact that money…how to manage it, how to create it, what to do with it…is often a taboo subject. If we don't talk about what to do with it, he points out, then we don't how to control it, we can't change our behaviors around it, and we are influenced *by* money rather than the other way around.

The same is true for veteran mental health.

For example, one study looked at over 150 primary care physicians in Rochester, NY. The focus of the study was to determine how often primary care physicians explore the topic of suicide[60]. The researchers discovered that the doctors explored suicide only 36% of the time when seeing clients with depressive symptoms. This is happening while the CDC reports[61] that nearly every state in America has seen a dramatic rise in suicide rates. Does this mean that the doctors don't care? No, it's more likely because suicide is something *we just don't talk about.*

I hear it a lot from clients that I work with. "I haven't talked about that in years" or "I don't think I've ever told anyone that." The problems that we're trying to solve won't be addressed if we don't talk about it.

Not Talking is Avoiding

One of the "problems" when it comes to talking about things is that we then have to acknowledge that they exist. Sometimes the things that we don't want to talk about are things that we don't want to admit to ourselves. We have a drinking problem, maybe, or I don't really feel satisfied with my current job. I don't want to recognize that my marriage is falling apart. If we don't talk about it, then maybe we can sort of just pretend it's not there.

Information avoiding is a way to cope with distressing news, but not an effective one. Recent studies show[62] that individuals will avoid information about an illness or their risk for disease if they're not comfortable with hearing it. Not talking about it leads to greater consequences than the distress that comes from avoiding the topic, but we still do it. How often do we "go along to get along" in our workplace or our relationships?

Not Talking is Not Defining the Problem

A commonly used form of psychotherapy is Cognitive Behavioral

Therapy (CBT). CBT is an intervention that helps individuals modify dysfunctional thoughts, emotions, and behaviors. I have found that veterans especially respond well to CBT; there's a system, there's a step-by-step process to challenging our thoughts. The first step, though, is awareness[63]. If we don't understand the shape of what we're dealing with, then we're not going to be able to impact it in any way.

One of the key elements of CBT is challenging false negative beliefs. Something like "I'm not worth their time" or "I'm fundamentally unlikeable." We may be thinking it all the time, but until we *talk* about it, it doesn't become "real." I've heard veterans say, "Well, now that I say it like that, it doesn't make a lot of sense. It did in my head, though." If we can't describe a problem, then we can't solve the problem.

Having Unspoken Conversations is Not Communicating

How many times have we thought we were in agreement with someone about a topic, but it turns out that we weren't? We just kind of assumed that they saw things our way, and they assumed that we saw things their way. Even the closest relationships have off-limits topics. Sometimes this is beneficial, but other times it isn't. Topic avoidance certainly has a place in our relationships. Depending on the situation, certain things are simply not talked about. We don't openly address medical or psychological concerns. Employment advice[64] identifies three taboo topics in the workplace: sex, politics, or religion. Agreeing to not talk about certain topics[65] helps us manage the tension between our need for openness with others and our need for privacy and protection.

Sometimes we can take that too far in personal relationships. We have so many off-limits topics that the things we *can* talk about are limited to sports and the weather. We pick our way through distressing conversations like tip-toeing through a minefield; or, more likely, avoid communicating at all.

Limiting Disclosure

Like with everything else in life, there needs to be balance in what and how much we talk about. The problem here is that we don't talk about *anything* when it comes to veteran mental health, but that doesn't mean that we have to talk about *everything*. Finding someone that you feel like you CAN talk about anything with is important, while recognizing that you don't have to tell them everything. Talking about things with others, though, can help to uncover the subjects we're hiding from ourselves.

Once we talk about these unspoken things, then we can control how we think about them, influence our behavior around them, and change for the better. And that's never a bad thing.

Hey, Veterans: Why So Serious?

Hey, defending the safety of our nation is serious business. We also know, of course, that not all of our military service was Rambo; there were days of mind numbing boredom. Senseless tasks. Police call, hurry-up-and-wait and a thousand minor annoyances that the movies don't show. If you've never served, you may think that your monthly company meeting is pointless, but when I was in the Army, I literally had meetings to plan for meetings that would plan for future meetings. It's like staring down hall of mirrors.

After leaving the service, the seriousness doesn't end. I'm a mental health counselor, so of course my job is serious. It's serious business. Check out my podcast episodes: they're about topics like veteran suicide, a lack of purpose and meaning, PTSD. Heavy, important subjects. And I'm always looking for self-improvement, process improvement, reading great books by inspiring leaders. A lot of the things we talk about on social media carry the weight of seriousness...what protests mean, what it means to serve, the politics, the idea that we have to express our very important and necessary opinion...over and over. It's all so...serious.

All of this seriousness isn't great. It's not good for stress, mental health, physical health, any of it. As good ol' Jack Torrence says in The Shining, "All work and no play makes Jack a dull boy." Okay, so maybe not the best reference regarding mental health, since he turns into a crazed attempted murderer, but you get my point. All seriousness and no fun DOES make us into something we don't want to be. We *do* life, but are we truly *enjoying* life?

Bring Back Dumb Memories

We all have them. The pranks we played on each other. The stories from your service that made you laugh then, and made you laugh now. Anyone who has deployed to Iraq or Afghanistan has stories about the strange way that animals were transported around the community: tied to the back of buses, shoved in trunks. I was once on a patrol in Nangarhar province in which I saw two goats tied to the top of a minivan. They were laying down, tied down by ropes through the windows of the vehicle; I was certain they were dead. I was staring out my window, and as we passed the minivan, one of the goats *lifted up it's head and looked at me!* That kind of thing happened *all the time*...startled me, sure, but it was also funny, looking back on it. Maybe there were times that the only choices you had were to laugh or scream, so you chose to laugh.

Regardless of why or how, when thinking back on your service, don't just focus on the painful stuff...think of the dumb stuff. The weird things that you bring back with you, the quirks that are part of your personality now that are directly tied to what happened when you were in the military.

Strangely, for me, a lot of it has to do with coffee. My wife can't stand it, but I recycle coffee grounds. I'll take a fresh scoop and throw it on top of the old stuff...this morning's grounds, yesterdays, doesn't matter. Drives her nuts. But that's what we did when I was a Platoon Sergeant in Afghanistan; you didn't care how it tasted, only how it did it's job. You had to ration that stuff, you know, and make it last for when you need it. Remembering stuff like that, for you, brings back an echo of that brotherhood, sense of goofiness, the levity that occurs in even the most stressful times. You weren't so serious all the time then...why are you now?

Do (or Watch or Read) Something Silly

Again, this is something that I realized recently. I wasn't reading anything for enjoyment, I was reading for *work*. I love my work, of course, and it's great; but I began to realize that all of my "leisure" time was being consumed by work. Even if I play games on my phone, or something like that, I'm still doing something serious; being distracted from thinking about work, or what I am doing tomorrow. Editing something, writing something, recording something, listening to something.

When you find yourself being so *serious,* then maybe do something simply for the humor or joy it brings you. I like old Marx Brothers movies; have them all saved. Put on the Three Stooges, if that's your thing. Old Bugs Bunny cartoons, or whatever it is that you enjoy. Watch something *silly,* not something *serious.* There is enough drama in our lives for us to seek out more to engage in.

Relax...Seriously

I've said it before, but there was often a rest plan[66] when we were in the Army, but we can also be consumed by the *need to keep going* in our post-military lives. That's the path to burnout. In 2011, I was between deployments to Afghanistan. I had just finished a tour of duty as a Company First Sergeant, and was preparing to make a move to Fort Polk, LA in order to deploy to Afghanistan on a Mobile Assessment Team for the Afghan Ministry of Defense.

I turned my company over on a Thursday, and on Friday my family and I had packed up the car to spend a week at the North Rim of the Grand Canyon. If you've never been to the North Rim, it's a pretty wide-open space...no cell service, and no paved roads, at least that we saw. We were legitimately and entirely unplugged for a week. I had books to read...about Afghanistan, as a matter of fact, and it turned out to be the book Horse Soldiers, the basis of the movie 12 Strong...but it was a real live book that I held in my hand, not on a device. More importantly, that was a time of *relaxation*. Personally, I believe that if I didn't unplug like that, where there was no communication between me and my unit, I would not have had as easy of a transition.

The same can be said in our post-military lives. Running ourselves into the ground is not beneficial for anyone…us, our family, our team, any of it. By all means, do serious work…but don't take it too seriously.

PART 4

Who We Will Become: Moving Forward in Post-Military Life

Several years ago, a colleague working with me as part of our local veteran's court asked me, "how long have you been out?" He was a retired Army First Sergeant and a correctional officer with the El Paso County Sherriff's Office. At the time, it was something like three years or so; I told him and he said, "wow, it seems like it's been a great transition for you...you've done well for yourself." I was surprised and taken aback...it sure didn't feel like it was great from where I was standing.

In a similar experience, I was having lunch with a mentor who asked me the same thing. He was a Navy veteran who had three tours in Vietnam...so he was definitely doing some hardcore stuff there...and then became, in succession, a police officer, a Presbyterian minister, and finally a couples counselor. When he asked me the question, I responded, "It's been two whole years!" He laughed and said, "kid, you still have sand behind your ears. Come talk to me when you've been out fifteen years."

The thing about the future is that we don't know what's out there. Maybe we want to cling on to the past because it's what we know, no matter how much it sucked. Maybe we're too overwhelmed with simply surviving in the present to think about what things will be like for us in twenty years. The fact is, however, that we will certainly be travelling down the Veteran road for a very long time.

As I write this, the WWII and Korean War generation is passing. There is not a Vietnam Veteran that is younger than 65; the service members who fought at the height of that war are in their seventies. The future is the unknown land where we're going to live, so we might as well start to get ready for it.

After the Military, Finding Your What Is As Important as Finding Your Why

March 1996. I was the lowest ranking guy in my platoon. My unit pushed south from our base in Hungary to Camp Angela, Bosnia. Being the lowest ranking guy in the platoon this didn't always mean that much, but this time, it meant I got the crap assignment. I was sent down early as a weapons guard to watch over our platoon's extra equipment: radios, extra weapons, stuff like that. They gave me a case of MREs, a case of water, and said, "see you in a few days." It was me, some other guys with the same crap detail, and about a hundred of my new Bosnian friends still building the camp around me. I remember, very clearly, sitting on the steps of my tent, looking at this huge hill across the quarry from me, wondering: "What the hell am I doing here?"

In many ways, and in many different places, I've asked that question, and many of the veterans that I work with asked themselves that question. Looking out the door of a perfectly good airplane (although it wasn't always perfectly good), pulling guard on some featureless expanse of desert, engaging in some futile exercise. The same questions: what and why.

I hear it from them after they get out of the military, too. "What's my purpose? What am I supposed to do now?" As I've often said, finding meaning and purpose[1] after leaving the military is significant to a successful transition Even those who have had a "successful" transition, though, can run up against this question: what is the purpose of my actions? What, really, is the problem I'm trying to solve? I may think that I'm trying to solve one problem...putting food on the table and shoes on my kids' feet...but I might be trying to solve an entirely different problem.

Sometimes the "What" May Be Something You Don't Like

We all have to do what we have to do. The problem you're trying to solve right now, hopefully, won't be the problem that will exist in the future, because once you solve it, it's done! Often, however, the problem we are solving is not one that we might want, but its the one we need to do. That week in Bosnia, the "what" I was doing was solving a problem, one that my leadership had: they needed to send extra equipment down to Bosnia and it couldn't be left unsecured. I was the solution to that problem. I didn't *want* to be the solution to that problem, because it created a problem for me, but I was the solution nevertheless. Really the worst of it was that I was bored out of my mind, but that's part of military life. After we leave the service, we may find a problem to solve that we may not enjoy; it doesn't have to be that way forever, but we didn't always love what we did in the military. We sucked it up then, we can suck it up now.

Sometimes There Doesn't Have To Be an "What"

I've seen this at work in my own life; I enjoy problem solving. I'm good at it. I'm so good at it that, if there is no problem, I might just create a problem so I can solve it. If we're used to operating in a chaotic environment, we might be a bit uncomfortable in an environment of peace...so we create chaos around us, so we can be comfortable. We don't always have to find problems to solve, though. The problem we might need to solve, the "what" we are looking for, might be our incessant problem solving. Recognizing that not everything needs to be figured out, not everything needs to be fixed, can be the solution to the problem we didn't know we had.

Maybe you think you know the "What" but are completely wrong

Leaving the military, I know that I was really anxious: I had to find a job. I had to find a job. There were job fairs I attended even before I dropped my retirement packet. I was leaning so far forward in the foxhole that I was falling out of it. I did lad a great job, working with an organization that was helping to house homeless veterans. To this day, I am grateful and appreciative of the time I spent with that organization; but the problem I was trying to solve (get a job) and the topic I wanted to focus on (veteran mental health) were nowhere near the same problem. I thought I knew the "what" I wanted to do: work with veterans. It turned out not to be the right "what." It was in the ballpark, and gave me valuable and appreciated experience, but it was someone else's "what" that I was trying to solve, not my own.

Find the "What" You Were Put On This Earth to Solve

Each of us has a unique set of experiences that gives us the ability to solve problems. To find our "what." The thing we did in the service doesn't need to be our "what," and in fact it rarely is. It is possible, and even necessary, to reinvent ourselves after the service. My father was a payroll clerk and courier in Vietnam. He became a cop after he left the service. I've known snipers who were really good at working with roofing companies, because scanning a roof for damage is pretty close to scanning your sector for something out of place. Maybe the problem you can solve is the lack of steady and reliable crew leaders for a moving company. Maybe it's in tech, or bioscience. Finding the problem you are uniquely qualified and able to solve can help you find that purpose and meaning we were talking about earlier.

Fall In Love With the "What," Not The "How"

This is a trap that we all fall into: being in love with our solution. The "How" is the solution, how we solve the "What" of our problem. If we are so stuck on the "how," we're going to use our comfortable old hammer to beat on everything because we think it's a nail. This goes along with the idea of carrying on with what you did when you were in the military; you enjoyed what you did (mostly) and you were good at it. Why not take that to your post military life and apply the same solution to the problems out here? That creates problems in and of itself, of course, because not all solutions match every problem.

When looking at "what next," take some time to consider what exactly the "what" is; you will be much more satisfied with what you're doing.

Getting Perspective in Post-Military Life

I'm no good at being noble, but it doesn't take much to see that the problems of three little people don't amount to a hill of beans in this crazy world – Rick Blaine, Casablanca

Sometimes, when we're in the middle of a firefight, it's hard to see the big picture. It's equally challenging to understand what's really going on if someone's watching the battle unfold from a camera in the sky. In post-military life, especially when it comes to veteran mental health, sometimes we can get caught up in our perspective of our problems. If we're isolating, we think that every veteran is experiencing the same thing. Without taking a step back or moving away from the situation, we can become overwhelmed by events and become stuck in a stagnant routine.

Getting perspective provides context and control to the situation.

In many of the situations that veterans find themselves outside of the military, it's easy to slip into a routine that makes you wonder where the week, or the year, went. When we were deployed, it was called "Groundfob Day," that seemingly endless routine that happens day in and day out and never seems to change. Get up, chow, patrol or work call, get back, go to the gym, eat, hit the rack. Over and over again. That can also occur in the workforce, or in school, and especially if we're not doing anything...just stuck in the rut of our post-military lives.

In other times, like when we're in an emotional or psychological crisis, the perspective can shrink to only that time period, that moment or that hour. Nothing mattered before this moment, nothing matters after this moment, we're just in the middle of it. It can be something that triggered our anxiety, or spun us into depression, but whatever happened, we're there. Even when time passes, we can get caught in that moment, angry that it happened, worried that it's going to happen again.

Get Perspective with the Passage of Time

With many things, the passage of time gives us a different point of view. Life looks very different at twenty compared to forty, and forty looks different from sixty. Some of that is life experience, some of it maturity, and some of it is just learning. As humans, we are unique in nature in that we have the ability to reflect on our past and predict our future, and reflecting on our past can give us greater perspective on our situations. In an article in 2013[2], Black and Liao indicate that this reflective nature is a survival mechanism. It is required to be able to orient ourselves to time and place...we are now and we are here...but also giving understanding and meaning to our lives.

A unique challenge for some veterans, however, is that traumatic memories resist reflection and do not seem to live in the past, but instead persist in the present. Bessel Van Der Kolk, a leading researcher on Posttraumatic Stress Disorder, observed that[3] while some memories fade and change over time, traumatic memories do not. Time or things that happen after the traumatic memory do not seem to impact the event; in this case, another way to shift perspective is necessary.

Get Perspective with an Outside Point of View

Have you ever had a conversation with a friend of family member, and they said something about your situation that you had never considered? Many times, if we're stuck in a place in our head and we're focused inward, we don't see what's right in front of us. I describe it to the veterans I work with as rubber bullets bouncing around in our heads[4]. If we allow it, these thoughts or judgements we have about ourselves and the world will get trapped in our mind. Letting them out in any way possible can be helpful. Sing it, shout it, draw it, write it, getting it out and into the real world in some way can allow us to step away from it.

Getting the point of view of another person, however, is a great way to get perspective. Maybe if we're in the middle of a battle, and we hear from someone who's looking at it from the camera in the sky. A mental health counselor can definitely support with that. Countless times, I've seen veterans who had not considered an alternate way of describing their situation. Bouncing your thoughts off of someone else is a great way to get perspective.

Get Perspective by Looking Around

Finally, we get so stuck in our heads that our world shrinks to us and our problems. Veterans are uniquely aware of the situation in the rest of the world, as they are much more likely to have been exposed to other cultures than those who haven't served. It's sometimes good to compare our current situation to that of others. This isn't some, "buck up, trooper, at least you don't have it as bad as them" statement. Instead, this is an accurate reflection on what's happening in the life of other people. This happened to me recently; I was caught up in something in my own head, and had breakfast with a colleague…then found that they had been mugged the night before. Talk about getting perspective really quick…my own problem paled in comparison. That "hill of beans" I'm talking about in the Casablanca quote.

If you're caught up in a storm, then take some time to get some perspective. Just looking at it from a different angle will provide the solution you're searching for.

On Hope and Veteran Mental Health

Hope. Those who have it, can't conceive of a life without it; those who don't have it can't dream of a life with it ever again. Hope is critical to our psychological wellbeing. If we have no expectations of good things in the future, and more damaging, stop wanting good things in our lives, then despair sets in. I've seen this often in the veterans I work with as a clinical mental health counselor. If they walk in full of pain but are hopeful, then there is a good chance that they can get to a point where the pain is not impacting their life as much. If they are in pain but have no hope, we can get there too...but it's going to take much longer.

Hopelessness and depression are linked. One can be depressed without hopelessness, and one can be hopeless without depression, but the two together can be devastating. And the combination of the two, and especially with hopelessness, means that the service member, veteran, or military family member are at a greater risk of suicide. One study looked at the impact of hopelessness as an exacerbating influence[5] on other life stressors. The researchers found that negative life events, such as homelessness, job loss, and bereavement, significantly increased the risk of suicide for the individual who was hopeless as well.

Meaning and Purpose is Critical to Hope

A critical part of a fulfilled post military life is that a veteran has something to do that satisfies them. We've talked about meaning and purpose before, but research shows that former service members who have a sense of purpose described being more hopeful about the future. On the other hand, veterans who feel like they're a burden expressed more hopelessness. When a veteran is trapped by hopelessness...that there is nowhere else to go or nothing else to try...then they are well and truly trapped.

Just because we feel like there's no hope, however, doesn't mean that there isn't any. This is the insidiousness of the betrayal of our own mind. When we get caught in a trap of hopelessness, it's as if we're at the bottom of a well. We don't enjoy it (meaning), don't feel like we can do anything about it (purpose), and can't see anything but stone walls in front of us regardless of what direction we look. Except...if we look up, and there's a way out. Someone who is hopeless will dismiss the idea that looking up is nothing but a feel-good cliché, and the person who is hopeful in their life recognizes that it's the truth.

Hope That Pain Will End

One benefit of hope is that it will get us to a place of relief if we're in pain. In this context, pain is any discomfort that we're feeling as a result of injury, illness, or negative life events. Physical pain, emotional pain, psychological pain, spiritual pain. Pain without the hope of relief is suffering, and suffering is unbearable[6] if we think it's not ever going to end. The problem happens when we're suffering and we can't think of a time when we won't be. And it becomes a recursive loop; again, as above, when it seems that when I try something, and it doesn't work, and I try another thing, and that doesn't work, then why try? And we fall back into the trap of pain that becomes suffering.

The thing is, I can be hopeful that someone will be free from pain. I often am. I see a veteran I'm working with who is suffering and hopeless, and I know they can get out of it...I've seen other veterans who feel exactly the same way regain their hopefulness and ease the pain. But I can't be hopeful *for* someone. I can't pour the hope I have for them into a cup and have them drink it. The hope must come from within the individual rather than from the outside...and so we're back to someone who doesn't have hope can't conceive of ever having it again.

Hope That Good Will Come

Another benefit of hope is that it will not only bring us relief, but joy and satisfaction. Mental health and wellness is not just about repairing deficits, getting us back up to baseline when we're down in that well. It's also about getting us above zero, having a life worth living. We can be free of pain and still not be loving life. The absence of pain does not mean the presence of joy. Hope can help us understand that life is better today than it was before, and life tomorrow can be better than life is today.

The one thing that can develop hope is perseverance. That's the one piece of advice that I can give: *don't give up*. You hear it all the time, and it drives pessimists up the wall: buck up, little trooper. The sun will come out tomorrow. The little engine that could. Perseverance is sometimes equated with blind pie-in-the-sky optimism, but that's not the case. It's more about knowing that, when we're in the middle of the long slog, that it will end. And it will, if we keep at it long enough.

Hope is critical to our wellbeing. It was like that when we were in the military...when we had that toxic leader or were stuck in some crappy field problem with mud up to our necks. There was hope that, eventually, this would end, and things would be better.

We can have the same thing in our post-military life. We just have to work to keep hope alive.

Who We Will Become: Moving Forward in Post-Military Life

Finding Flow in Post-Military Life

"Of all the virtues we can learn no trait is more useful, more essential for survival, and more likely to improve the quality of life than the ability to transform adversity into an enjoyable challenge." — Mihaly Csikszentmihalyi

Have you ever been so engaged in an activity that you lost track of time? You were so focused on what you were doing, you look up and realize that it's been three hours. You didn't even realize it.

That's a state of flow, or being in the zone. That state of mind where you were doing something you enjoy so much, and are so good at doing it, that you get totally lost in the experience. Flow can be achieved by as many different things as there are people; hiking, job tasks, childcare, you name it.

For many veterans, a lot of their time in the military was punctuated by these flow moments. Sports and physical activity can sure do it; going to the range, executing battle drills, teaching your subordinates how to execute a difficult task…one of the reasons that veterans enjoyed the military so much was the opportunity to engage in a flow state, and if you found something you loved and you were good at, then you were able to operate in flow pretty often.

There were times, of course, where flow was not in the picture. Think guard duty[7]; An endless repetition of uneventful shifts guarding an ammunition supply point that is in a base that is guarded by others. Or constant police call…picking up trash, for those who never served. Endless repetitions of inspecting vehicles for deficiencies. These are the times when we recall the mind-numbing boredom that went along with military service, and miss those moments much less than the flow moments.

It is possible to engineer flow moments in our lives, so that we can get the same satisfaction post military that we did while we were in.

The Concept of Flow

The concept of Flow was identified and popularized by Hungarian-born psychologist Mihaly Csikszentmihalyi (pronounced Me-High Chick-Sent-Me-High). It has been variously described, as I have above, as being lost in a task. Being so involved in what you're doing that you lose track of time. It is doing something just for the enjoyment or positive feelings you get from doing the task.

Csikszentmihalyi describes Flow in this way for an interview in *Wired*[8] magazine:

"…being completely involved in an activity for its own sake. The ego falls away. Time flies. Every action, movement, and thought follows inevitably from the previous one,

like playing jazz. Your whole being is involved, and you're using your skills to the utmost."

In a chapter on the concept of flow in the Handbook of Positive Psychology[9], Nakamura and Csikszentmihalyi identify six factors required for someone to engage in a state of flow:

Intense and focused concentration on the present moment

Merging of action and awareness

A lack of awareness or self-consciousness

A sense of personal control over the situation or activity

A distortion of time

Experience of the activity as personally rewarding, regardless of any external motivation

Key Elements of the Flow State

Two of the key elements of the flow state are how difficult a task is, compared to our skill level in accomplishing that task. Think way back to when we were learning to tie our shoelaces. Now, that task is extremely easy; it is so routine that you probably don't even remember doing it this morning. Back then, however, it was extremely difficult, and our skill at doing it wasn't very highly developed. Have you ever watched a kid trying to tie their shoes, the look of intense concentration, maybe the tongue poking out of the mouth a little bit? Very quickly, however, the task became easier and our skill became greater; the situation reversed. We were no longer in a state of flow. We became bored quickly; our skill was much greater than the difficulty of the task.

Consider the opposite scenario; the difficulty of the task is so great, and our skill at conducting the task is so low, that we become frustrated. I've heard veterans feel this way when they go back to school, and get into courses where they just don't get the material. Very quickly, they become irritated…at themselves, their classmates who seem to be getting this stuff easily, at their instructor. If a task is so challenging that we can't seem to figure it out, then we quickly lose interest in that as well.

Balancing Task Difficulty with Skill Ability

The key to engaging in flow is to find something that has a difficulty that is just above or just below your ability. As your ability increases, the difficulty increases. Woodworking, for example; we start off with small projects, just the right amount of difficulty, and increase the challenges of the pieces we're working on as our skill increases. This applies to many different tasks that we can do in our post-military life, both through employment and hobbies. Running: you start off with a 5k, increase your endurance and distance and eventually you're running a marathon. Hiking: start with the trails near the house and eventually end up on the mountaintop.

You hear stories all the time about veterans who get caught in jobs they hate after leaving the military. Many times, it's because these jobs don't offer the same opportunities to engage in flow states that the veteran had when they were in the military. When your skill...leadership, task accomplishment, quick and precise execution...is greater than the task at hand, then you get frustrated. Just getting the paycheck isn't enough...veterans need to have intrinsic motivation for the work they're doing, it has to be personally rewarding. If it isn't, boredom and apathy quickly sneaks in.

Engineer Flow Experiences in Post-Military Life

Once you understand the concept of flow, you can find things in your post-military life that get you there. It doesn't just have to be about work; work can be about the money, and other stuff you do can be about flow. Yard work. Taking care of the kids. Hobbies and outside activities. Keep in mind, these things should have both meaning and purpose...they should be both satisfying, and have tasks to be accomplished with an identifiable end result. You can certainly get into a flow state by playing video games for six hours...but have little to show for it at the end of that time.

If you engineer flow states, then you're more likely to experience satisfaction in your post-military life.

Veterans Can Find Meaning and Purpose through Service

The best way to find yourself is to lose yourself in the service of others – Mahatma Ghandi

One of the most common impacts of leaving the military is loss of meaning and purpose[10]. Regardless of why we joined, while we were in the military, service members expended energy and effort for the greater good. There was the global "service to the nation," of course, but service to those around us, our battle buddies, shipmates, or wingmen. To a greater purpose.

When a service member leaves the military and doesn't have that, things can go wrong quickly. We can become bored. Disillusioned. Angry even, and there is a danger of falling into the veteran stereotypes. A hero, to be revered; a victim, to be pitied; a villain, to be feared. Sometimes, veterans play into those stereotypes. We rage and rail against perceived injustices to anyone who will listen. To what purpose? What, precisely, are veterans trying to do when we shout "unfair" and "where's mine" to fill the void that leaving the military left in us? Some of it is to certainly fill that void, but it is ultimately unsatisfactory.

It's hard to shake your fist at the sky when you're holding a shovel.

There's plenty to do in our post-military lives, and to be sure, getting out of work is something that we often tried to do when we were in. The thing was, when we were shamming in the military[11], it was often busy work, and there were others to pick up the slack. And, because you can't be lucky enough to execute the proper sham maneuver every time, you often picked up the slack when others shammed. So it all worked out in the end. Not doing work outside the military, though? Although you might not truly be on your own, you're doing the solo thing just enough that there's not anyone to pick up the slack when you're shamming it.

As I mention often, the current generation of veterans has the ability, and responsibility, to change the world[12]. Talk doesn't bring about change, though, work does, and one of the best ways to fill that void is to work on behalf of others.

Service through Veteran Groups

"Who wants to join a bunch of old guys drinking in a basement?" I don't hear it a lot, but often enough that it's noticeable. The divide between Post 9/11 veterans and those who came before them is often identified by organizations that support one group over the other. We also divide ourselves in the same way. Younger veterans are staying away[13] from

145

"legacy" veteran service organizations like the VFW[14] and American Legion[15]. Just like the stereotypes that people have of veterans, however, we can sometimes buy into the stereotypes of these organizations. How do you know how they are, if you don't check it out for yourself? I don't belong to any, for full disclosure, but for me it's because it's a matter of *time*, not inclination.

Organizations such as these are doing great things on the part of veterans. Former soldiers of mine are working in organizations like the Vietnam Veterans of America[16], of the Paralyzed Veterans of America[17]. Others are getting involved in other veteran organizations that give back to the community, like Team Rubicon[18] or the Travis Manion Foundation[19]. If there is a loss of a sense of service and camaraderie in our post-military lives, getting involved in organizations such as these can help.

Service through Entrepreneurship

One of the best ways to do something is to do something. I know it sounds cliché, but building or creating something new can provide both satisfaction and service to the community. Organizations these days are not just concerned with making stuff and money, as it was in the old days. There is a sense of social responsibility that many organizations have, and that social responsibility can be provided both in large organizations and small organizations. Entrepreneurs are driven individuals who identify a problem or a need, and come up with a solution that solves that problem or satisfies that need. Sound like a veteran? Take a look at this list of entrepreneur definitions[20] and see if they meet the definition of a former military service member.

Don't know how to do it on your own? You don't have to. Amazing organizations like Bunker Labs or WeWork are supporting veteran entrepreneurship. Opportunities such as Stanford Ignite and Institute for Veterans and Military Families through Syracuse University are designed to help veterans launch their companies. The good ol' Department of Veterans Affairs even gets in on the act, providing a whole bunch of information for veterans looking to start their own business.

Service in Legislative Advocacy

We need more politicians like we need another hole in our head, right? And getting involved in legislative advocacy doesn't mean that you have to run for office. Just that you have to be involved in the process. It's not as hard as you might think; you just have to show up to talk to people. I believe one of the reasons why veterans don't get involved in the legislative process is because we're not used to being involved when we were in the military. Then, it was forbidden; after the military, it's encouraged. You never know what you can do if you reach out to lawmakers. I found myself at the Colorado State Capitol for the first time several years ago, wondering how I got there; now, while I'm not an expert, I've certainly done my part to make an impact.

Many national legislative advocacy efforts are made on behalf of the legacy VSOs, as mentioned above, but one organization making a difference at the national level is High Ground Veterans Advocacy[21]. And on the local level? Many might not know this, but state lawmakers don't have a full legislative staff and most have other careers throughout the year. They rely on interested constituents to get things done…which is a great opportunity for a veteran looking to make a difference.

Veterans have a lot to give in our post-military lives…we just have to find a mission and go for it.

Being Grateful for the Least of Things

If the only prayer you ever say in your entire life is thank you, it will be enough.
Meister Eckhart

Gratitude has the power to uplift the heart and raise the spirit. I've written before about the power of gratitude[22], but it is always good to take the time to reflect about what we're grateful for. For many veterans, gratitude is focused on their coming home from some of the most significant challenges a human can face; their families, even more so. This is not universal, of course; many veterans feel guilt instead of gratitude, and this can hold them back. Is it possible to be both thankful for our own life, and have remorse for loss? Sure, but some veterans think the two are mutually exclusive.

Whenever I think of gratitude, I think of the traditional Hebrew song, Dayenu, which is sung to celebrate the Passover Seder. To be perfectly clear, I'm not Jewish, but I don't have to be in order to appreciate the wisdom, humility, and gratitude in this concept.

The term "Dayenu" means[23] "it would have been enough." The term is sung as a chorus after a recitation of a series of historical events, starting with the Israelites being rescued from slavery. The concept goes like this: "If only we would have been rescued from slavery, but not carried out judgements against our oppressors, it would have been enough. If He had given us their wealth, and had not split the sea for us, it would have been enough." Basically, the concept of Dayenu means that even the least amount of thing that we are blessed with is enough, and anything beyond that is even a greater blessing.

Being Grateful for the Least Benefit

So how does that apply to our lives? By considering the concept of Dayenu, we can literally count our blessings, reflect on those things that have happened to us, and appreciate them individually. For example: if I had met my wife, but not had children, it would have been enough. If I had children, but never get grandchildren, it would have been enough.

This is an exercise for focusing on the thankfulness of the least thing and seeing everything beyond that as an even greater object of appreciation. We can focus on small benefits, and build that gratitude on top of gratitude. It is considering our blessings individually, starting at the first, and appreciating them all at the level that accumulates into greater appreciation.

Not Being Grateful for Grateful's Sake

This is not simply saying "Well, at least I..." as in "At least I have all my fingers and toes. What do I have to complain about?" It goes deeper than that: if the only thing that I had was my fingers and toes, then that would be enough; but then you go beyond that. Yes, I have my fingers and toes…and my health. If the change stopped there, that would be a great blessing. But then we keep moving up the chain even more: if I came back from combat, it would have been enough. If I came back from combat and returned to a stable family life, that would have been enough. If I came back from combat, returned to a stable family life, and got a meaningful job after leaving the military, *that* would have been enough. This exercise is about starting with the least best thing, and heaping more gratitude on it from there.

You Have To Believe What You're Trying To Be Grateful For

This isn't a reverse psychology mind trick. You have to be truly grateful for what you're thinking about, and mean it. I retired from the Army as a Sergeant First Class; of course I wanted to retire at a higher rank, but circumstances didn't allow it. If I say, "if I had been promoted to E-7, and not been able to lead a platoon, it would have been enough," then I have to *mean* it. There were times in the past where I *didn't* mean it. There were times that I was bitter about being passed over for promotion, for spending more time in the military as a Sergeant First Class than I did all other ranks combined. However…in spite of my rank, I was given the responsibility of being a company First Sergeant. So now, in reflection, it goes like this: if I had been promoted to SFC, and not given the responsibility to lead a platoon in combat, it would have been enough. If I was given the responsibility to lead a platoon in combat, but not given the responsibility of running a company, *that* would have been enough. And so on…

Being Grateful for the Least Best Thing Increases Gratefulness

In this way, gratitude builds on gratitude. The accumulation is not additive, it's exponential; spending the time every day to reflect on our blessings increases our ability to think, see, and feel positively. It's important to understand that our outlook drives our mindset and our behaviors; how we see is what we feel. What are you grateful for, least to most? I'd love to hear it, and share in gratitude with you. Post in the comments below, or share a comment on social media.

If only one person would have read the words that I have written, and not been touched, it would have been enough. If one person was moved to action by something I've written, but not many people, it would have been enough. My gratitude, however, extends to each of you reading this, whether it is one or many…there's somebody in the world that gives a crap. And where there is one, there's more. The one is enough.

Stepping Out in Spite of Doubt

There are no constraints to the human mind, no walls around the human spirit, no barriers to our progress except those we ourselves erect — Ronald Regan

I am easily inspired by ideas. Whenever I hear something interesting, exciting, motivating, it's like a seed is planted in the fertile soil of my soul. It's always been like this, I think; I'm always on the lookout for the new and novel. I'm interested in a theory or a thought or an action that can improve.

Some say that I'm a dreamer.

I'm also a doer, though. A great idea is worth little if it's poorly executed, and it's worth nothing if it's not executed at all. Too many times, our own fears, our own doubts, can get in the way of our accomplishments. We find ourselves stuck in a cage of our own construction[24], trapped by feelings of doubt. There's one thing that I've noticed, both in my military career and in life after the military. Many veterans let their doubt hold them back.

We never know what we'll achieve if we take a shot. Step out in the face of doubt. Move forward with bold and decisive action. How often have you wanted to do something, but didn't think you had it in you...but when you did it, you found that you DID have it in you? That's what being in the military was all about, stretching ourselves to the limits of our perceived ability, and knowing that there was much more there.

It All Begins with One Thing

I once had a conversation with a veteran who wanted to write a book. They weren't certain they could do it, though; it was beyond their ability. The thing was, this veteran also ran marathons. What's the difference between running a marathon and writing a book? Both begin with one thing...for a marathon, a single step, for a book, a single word...and eventually, if you persist long enough, the thing is complete. The veteran was firmly convinced, however, that it was beyond them, so I gave them a simple task...just a thousand words on the topic of their choice, and I'd help them with it.

I'm still waiting for those thousand words.

Don't Remove Doubt, Advance in Spite of It

Moving forward in bold and decisive action does not mean not having doubt. There's always going to be uncertainty about things that we've never done before; the step between becoming a dreamer and a doer is to move out in spite of that doubt. What's the worst that can happen if you reach for something and it doesn't happen? You don't get it? Okay. It wasn't meant to be, pick up and move on. Do you doubt that you can get into Yale or Stanford? Perfectly understandable if no one in your family has ever been to college, much less some of the most prestigious universities in the country. But does your doubt make it true? Organizations like Service to School[25] exist to help service members do that exact thing…gain admission to the best college or graduate programs possible.

As Patton said, "a good plan violently executed now is better than a perfect plan executed next week." If we wait until all conditions are perfect to put our ideas into place, then those ideas will continue to sit on the shelf.

Let Them Tell You No

One thing I often hear from veterans is that they don't want to get turned down, they don't think they can handle the rejection. If that's what we believe, then that's what's going to be true. If we turn it around, though, and think that we CAN handle the rejection, that if the answer we get back is "no" and we're as okay with it as if the answer was "yes," then we're more likely to get a yes in the future. Or at least not lose out on the chance to receive a yes.

One of the best things that happened to me was to receive a scholarship in 2015. Not because it gave me money, which allowed me to purchase the nifty machine I'm using to write this, but because it gave me a mentor. Every single measure of my success, in my opinion, is due directly to my mentor, Dannette Patterson. The scholarship was for veterans who were studying to be mental health professionals. Seems like a shoo-in, right?

Except that I was turned down for the scholarship in 2014. And I had applied to be a Tillman Scholar before that, and not selected. If I told myself no…that there was no way that I would be selected as a Tillman Scholar (which was true, or at least it was that year), or that if they told me no once, they'll tell me no again, then I wouldn't be where I am today.

Test the Limits of What's Possible

We will never know what we can achieve until we try. There are some things that we're certain we can do, and they're within the limits of our ability. But what could we achieve through an audacity of the imagination? If we were bold enough to dream, plan, and then do? Testing the limits of what's possible will help us achieve what we thought was impossible. It doesn't have to be anything huge, but it does have to be something; and if we can get one crazy idea to work out, then the dream becomes reality.

The Role of Shared Background Knowledge in Post-Military Life

Veterans face many challenges in post-military life. It's not that these challenges can't be overcome, or that veterans are somehow deficient because they experience these challenges. It's what happens when we move from one culture to another. One of the challenges that frustrate many of the veterans I see is that others don't "get them." Others don't understand how they think, why they do what they do. Often the only ones who do "get them" are others who served in the military. Therefore, many veterans will avoid connecting with others, for the very reason of avoiding the discomfort.

One of the reasons for the frustration is that it's true; there are things that a former service member knows that someone who never served doesn't know. Just as there are things that you learn if you grow up on a farm that you don't know if you grow up in the suburbs or the city. The difference is that the city kid doesn't feel like the farm kid *should* know what they know. It's recognized that they come from different cultures, different backgrounds, and know different things.

Veterans, on the other hand, really do what others to understand what they went through. At the same time, we don't know how to tell them. As I've said elsewhere[26], this is the paradox of the veteran story: we want people to know what we did and how we did it, and nothing in the world can drag our story out of us. One difficulty is that veterans often assume that others know what we know, and when we have to fill in the little details in order to get to the bigger story, then there is frustration.

Shared Background Knowledge

In the field of education, there is a concept called "shared background knowledge[27]" that is critical to learning. A person's background knowledge is the accumulation of their experiences. Once acquired, that knowledge is used to inform any current information being processed. I heard an example recently. Take the phrase, "Polly put the kettle on the stove so she could make herself some tea." In order for that sentence to make sense, the reader or listener would have to know certain things. What kettles and stoves are, how you use them together to make tea, that Polly is traditionally a female name. All of these bits of information are the "background knowledge" that helps us to interpret the information presented.

Shared Military Background Knowledge

Anyone who served in the military shares a base of background knowledge with those who served. Each of the military branches, furthermore, have a base of background knowledge that is unique to their community and culture. These fall into the "everyone knows" category; when I try to describe the good and bad places to be stationed in the Army, I compare between Fort Carson, Colorado and Fort Polk, Louisiana. For the Air Force, we talk about the differences between Minot, North Dakota and Colorado Springs, Colorado. For Marines: the comparison of Twenty-Nine Palms to just about anywhere else in the world.

Those who served in the military develop this background knowledge that is shared by others with the same experiences, just as the city kid and country kid does. The difference is that this knowledge is laid on top of their previous knowledge, and each influence the other. This is one of the reasons for the "you've changed" comments that many veterans get upon returning home. We learned different things, we now have background knowledge that is not shared by those around us.

Shared Cultural Background Knowledge

The same is true going the opposite way. When a veteran enters a different environment...the workforce or higher education, for example...peers their same age (or younger) may seem to be ahead of them because of a lack of shared background knowledge. "They" seem to know things that we don't. The stuff we struggle with comes easy to them. The new terms, the strange concepts, all of these are challenging because veterans have no frame of reference for them. I'd like to see them break down and reassemble a rifle in less than two minutes, though...right? Some knowledge that veterans have doesn't seem to quite fit in post-military life.

One of the problems, and I hear it often, is that we are trained to be service members but nobody takes the time to train us to not be service members. When we joined the military, the shared background knowledge was drilled into us (no pun intended, Drill Sergeant). When we leave the military, though, we have to obtain new background knowledge ourselves. Many times, we don't know where to start, so we don't even begin...and feel the knowledge gap between those who served and those who didn't grow wider and wider.

Assumptions About Shared Background Knowledge

The true frustration occurs when veterans assume that others have the same background knowledge that we do, but they don't. Or when we think they "should" know what we know, often without us telling them. On the other hand, veterans get frustrated when others assume we know stuff that we don't. Understanding shared background knowledge goes both ways.

One of the ways to develop background knowledge, both in ourselves and others, is to be open to learning. To listen to each other, to share experiences. I have a colleague who is a mental health professional but is not a veteran. She told me once that her neighbor, a retired Master Sergeant in the Army, is often her source for understanding about the military. She doesn't require her clients to teach her about the military, she learns it from other sources...many of which include veterans themselves.

Once we both teach and develop background knowledge, then post-military life gets a lot easier.

The Importance of Social Connection in Veteran Mental Health

Life is not a solo act. It's a huge collaboration, and we all need to assemble around us the people who care about us and support us in times of strife – Tim Gunn

Humans are social creatures. Sure, there are varying degrees of introversion and extroversion that people are comfortable with. We still need other people in our lives to feel complete, though. When it comes to health and wellness in post-military life, the presence or absence of a social network…big or small…can make a huge difference in life satisfaction.

While social connectedness heals, isolation hurts. Think about it; in the deepest, darkest moments that we have experienced, were you alone, or were others there? And you don't have to be physically alone; many veterans feel like they're alone in the middle of a crowd. Isolation is as much a state of mind as it is a physical state.

Studies have shown that a strong social network, including family, friends, trusted colleagues, and positive social interactions, lead to greater life satisfaction in post-military life. Some of the benefit of the social network is that it can prevent a crisis. Another benefit is that it can provide support during a crisis. With a strong social network, a veteran may not ever get to the point of psychological crisis, and if they do, the network is there to support them. Our relationships matter, and without them, we're in danger. That doesn't mean to say that we should immediately turn into social butterflies; some of us aren't built like that. But without *some* type of social network, post-military life can be much more challenging than it needs to be.

Social Support Can Prevent A Crisis

Often, the greatest benefit that a social network brings is the ability to prevent a crisis moment before it happens. Here, I'm talking about a psychological crisis; a "nervous breakdown," or suicidal ideation. Blowing up, anger and rage. In one research study[28] that interviewed a range of military veterans from many different eras, having a strong social support was a protective factor against these types of incidents. Those former service members who had a lot of social connections reported better overall wellness and used less mental health services than those who were socially isolated. The social connections served as a sort of barrier between the veteran and the effects of combat, adverse childhood experience, and current life stressors.

The presence of a strong social network provides the veteran with more resources to avoid a crisis situation. Having someone that you can easily talk to, who supports us in a non-judgmental way, can reduce much of the stress that builds up and sends us to a crisis point. Just being able to reach out and know that there's someone there can reduce a significant burden. During a recent conversation with Sally Spencer-Thomas on the Head Space and Timing Podcast, she talked about having that "three in the morning friend." The one person, or group of people, that you can call at three in the morning and you know that they will pick up the phone; just knowing they're there can keep us from getting to a crisis point.

Social Support Can Alleviate A Crisis

Sometimes, the lift is too heavy, and we find ourselves in a crisis moment despite our social support. Further research helps us understand that having a strong support network can lead to better outcomes in treatment. In the same study described above, veterans who actively sought help from their support network were more likely to need and use mental health services. If a veteran is struggling with problems related to their military service, the help that they get from their informal care network leads to better outcomes in treatment.

This finding is supported in other studies, as well. In January of 2018, the National Academies of Sciences, Engineering and Medicine released a report on a congressionally mandated evaluation of VA Mental Health Services[29]. One of the most significant findings was that Post 9/11 veterans who have a strong social network that support the veteran seeking treatment were more likely to use mental health services at the VA than veterans without such support. I see this nearly daily with the veterans I see in therapy; those who have family and friends who support them seeking treatment report more satisfaction. Those that don't have the same support can struggle to improve.

Isolation Can Exacerbate a Crisis

And then there's the opposite side, the dark side. Being isolated, feeling like we don't have good social connections, can lead to a crisis moment. A veteran without a strong social network can find themselves isolated and alone. The Interpersonal-Psychological Theory of Suicide proposes that a suicidal crisis occurs when a veteran feels as though they are a burden to others, experience social isolation, and have acquired the ability to inflict-self harm.

A review of suicides[30] in the military by Carl Castro and Sarah Kintzle identified that, when a veteran feels like a burden to those around them, they are more likely to turn inwards. Without a sense of social belongness, this sense of burdensomeness can't be denied. And neither of these two factors would lead to self-injury if the individual hadn't acquired the ability to harm themselves. These three factors are like three parts of an equation; if all are present, there is significant danger. If any one of the three are absent, the danger is much less.

Social Networks Are Critical to Health and Wellness

The fact is, as I said in the beginning, humans are social creatures. When we served, the military handed our social network to us. For better or worse, the buddies that we had were the buddies we had. We may have hated some of them, and loved others, but at least they were ours. In post-military life, one of the needs we have to meet is building another social network. It needs to be as strong or stronger than the one we had in the military.

Satisfaction in post-military life depends on it.

Public Support for Veterans and Veteran Mental Health

On mid-tour leave from my first tour in Afghanistan, I was having lunch in an airport with a First Sergeant. I have no clue who he was; one of those connections you make on a military flight from one location to another. We start to shoot the breeze on the plane, then he says, "hey, let's grab chow." So we're in an airport restaurant…Chili's, maybe…and at the end of the meal, our waitress comes up and says, "Your check's been paid for. They said to say thank you." Pretty neat…and generous, considering that we were a couple of grown men eating state-side cooking for the first time in eight months or so. At airport restaurant prices, too.

I wrote recently about how the public perception of veterans[31] can impact veteran mental health. The constant, and seemingly disingenuous, "thank you for your service." How the community seems to love veterans from a distance, but change when the veteran gets close. Attitudes in the workplace surrounding veterans. And I stand by that article; all of those things are true, and according the comments I've received in reaction to it, many veterans think so too.

I also realized, however, that the article was pretty negative. I even called it a rant at the end. Sure, it was showing the reality of what many veterans, including myself, have faced. But it could give the impression that this is the only thing that veterans experience, which would be inaccurate. There's another side to the story, too.

Employers Who Truly Want to Know

I've been with my agency for just over five years. My two immediate supervisors in that time never served in the military, and never made me feel unwelcome. As a matter of fact, in many ways, my military service was the REASON I was brought on board. We had a program serving veterans in the criminal justice system, and the significant majority of them were male combat veterans. My military experience combined with my clinical training was an asset in my workplace; I know this is true for many veterans. Veterans on Wall Street[32] is a great example of the corporate world coming together to support veterans and their families; it's not about filling a quota or getting the good PR from saying, "we hire vets." It is representative of a genuine desire to give back to those who served.

Providers Who Genuinely Do Care

I have heard from many veterans who have said, "I can't talk to a therapist who hasn't served." I even heard of one veteran who said that they would not go see a therapist who hadn't been an Army Infantryman who deployed to Afghanistan during the height of the combat operations. These are barriers that we put in place ourselves. The fact is, there is an extreme shortage of mental health professionals in this country. If we start to get choosy like that, then we're limiting our options…and, perhaps, therefore avoiding therapy like we really want to anyway.

I have a number of dedicated colleagues who never served in the military, who never grew up in a military family, and yet have dedicated their lives to serving veterans. I recently recorded a podcast with a psychologist who has been working with veterans for over forty years, Dr. Larry Decker[33]. Larry was very much not connected to the military; as he said, he protested *against* the Vietnam war, and several years later he found himself working as a therapist with those who served. I have spoken to dozens of colleagues who get it, who understand what they don't know, but work hard to figure it out.

Community Members Who Go Beyond "TYFYS"

Not all of our neighbors who haven't served are clueless zombies mumbling "thank you for your service" and then proceeding directly to "did you ever kill anybody." Thinking that will turn into us treating all of our neighbors in this way, therefore contributing to the civilian-military divide[34]. I have known many individuals, both through their interaction with the Head Space and Timing blog and in my day-to-day life, who value and appreciate the sacrifice of military service members. A comment by[35] a huge supporter of the blog and podcast, Nelson Ormsby, the son and father of service members, but who never served himself, describes his own aversion to "thank you for your service."

So as an adult, walking to my office at BWI Airport past the Military Airlift Command or the USO post-9/11, instead I'd ask, "you headed downrange or coming back?", and if the latter, I'd say "it's good to have you home."

…After once welcoming a Trooper home, he disarmingly allowed in return "thanks, but my buddy didn't". I then asked if maybe he did come home because, by referencing him in the response to me, his brother clearly remained in his head and his heart: "Sure sounds like you brought him home to me". Then he talked and I listened.

That's the kind of interaction that warms this old Sergeant's heart. And that's not an isolated incident; there are many, many community members who never served who sincerely appreciate those who served in the military. Not in a patriotic its-my-duty kind of way, but in a genuine, heartfelt way. A friend, Aaron Skogen, is an example of someone who never served, and shared his growing appreciation for those who have in this guest post[36].

You Carry What You Bring With You

Like with anything else, what you see is what you will find. If you only see neighbors as clueless automatons who have no understanding or appreciation of your service, then that's what you'll find. You'll be missing out on a bunch of great conversations with people who genuinely do care. Similarly, if you only think that everyone who hasn't served is grateful and appreciative, then you're going to be sorely mistaken. Understand: you're complicated. People are complicated. Connect with those who care, distance yourself from those who don't, and carry on with the mission of finding peace and happiness in post-military life.

National Vietnam War Veterans Day - Until They All Come Home

Those who cannot remember the past are condemned to repeat it – George Santayana

Vietnam. The country, the war, the service members who fought in it. The word conjures up images, ones that are rarely pleasant, and ones whose impact is long lasting. And the recognition of those who fought, and died, in this conflict should never be forgotten and should always be honored.

March 29th is National Vietnam War Veterans Day. It was designated as Vietnam Veterans Day in 2012 by President Barak Obama, and formalized as an official holiday in 2017 by President Donald Trump by the signing of the Vietnam War Veterans Recognition Act of 2017. The significance of this date is that the last combat troops were withdrawn from Vietnam, ending our nation's direct military involvement in the war.

Vietnam, the war, was politically divisive and had a multi-generational impact. The war not only made it's mark on the nation, but on the individuals who served in the conflict, their families, and their communities, and the impact continues to be felt to this day.

While significant in many ways, for me, the impact of this war is personal. The echoes have influenced my life for as long as I can remember. As I've mentioned elsewhere[37], my father and three of my uncles served in Vietnam, with each of them having a different reaction in their post-military life. That war, my father's reaction to it, and the nation's reaction to those who fought in it had as much an impact on my life as my own combat deployments. In the most recent years, my father's service had come to be more respected and accepted. The importance of the designation of National Vietnam War Veterans Day last year, while appreciated, was not significant, to me, or even to my father; he had finally come to a measure of peace about it.

Then, in June of 2017, we lost him to a combination of natural causes and elemental exposure. And the opportunity to tell him how much I appreciated his sacrifice and honored his service was lost forever.

Make no mistake, the generation of men and women who served in Vietnam is slowly starting to fade away. As I visit my father's headstone in Fort Logan National Cemetery, I look around at the memorial markers of his fellow service members, and he is surrounded by Vietnam Veterans. There are a few Korean War veterans here and there, and one Gulf War veteran, but he is flanked on both sides by those he fought with. The impact of the passing of this generation is not being noticed, and should not be underestimated.

My father never talked much about Vietnam. That was how he reacted to it; we knew it was there, we saw the impact that it had on his life, but we didn't know much about it. I didn't fully understand it, as my parents divorced shortly after I was born, and I only saw the impact in bits and pieces. I recall when attitudes towards Vietnam Veterans started to shift, though. In the mid-80s, I attended a Welcome Home event with my father and my uncle. The Traveling Wall[38] made a stop outside of St. Louis, and I visited the Wall with them. Later that day, there was a huge barbecue and welcome home party; my father later mentioned that it was the first time he felt truly recognized and thanked for his service.

Later, when I was stationed at Fort Bragg, I offered to get him a rubbing of a friend's name from the wall; he appreciated it, but declined, and instead suggested that I get the name of a friend of my uncle's and give it to him. That was my recollection of his reaction to Vietnam: pointing towards others, minimizing his own service. Pretending that it did not have the impact on his life, and our lives, that it did.

It was only before my first combat deployment to Iraq that we had a serious conversation about the impact of war. The piece of advice he gave me at that time: "When you get back, talk to your wife about it. Don't shut her out and try to keep it to yourself. That was one of the places where I went wrong with your mother." I know that it wasn't the only thing that went wrong, but it was a big part of it.

During my retirement ceremony in May of 2014, the Colonel presiding the ceremony was reading the accomplishments and achievements of those of us who were retiring. As he was talking about my service, he mentioned that my father, a Vietnam Veteran, was in attendance. He stopped his prepared remarks, looked at my father in the audience, and said, "I want to personally thank you for your service as well, Sir. The soldiers of your generation paid a heavy price so that the soldiers of our generation can succeed." After the ceremony, a line of Colonels and Sergeants Major lined up to shake his hand and thank him for his service.

That recognition, while appreciated and honored by both me and my father, was forty-five years overdue.

The impact of the Vietnam war is still being felt today. The support that veterans of my generation are receiving from the Department of Veterans Affairs, as challenging as it can be sometimes, is a direct result of the veterans of my father's generation who refuse to allow what happened to them happen to us.

We are truly standing on the shoulders of giants. We are thankful for it, and should never forget.

The Silent Plea of Monuments

We have come to dedicate a portion of that field, as a final resting place for those who here gave their lives that that nation might live. It is altogether fitting and proper that we should do this.

These monuments of stone and steel cry out for remembrance. They call to us: remember the fallen. Remember the sacrifice. Remember those who gave everything, even up to their very lives. Remember those who answered the call, some willingly, some grudgingly, but all faithfully.

As we approach Memorial Day, I am reminded of the Gettysburg Address. We had to memorize it in elementary school, we know the stories of President Lincoln writing it on a top hat on the train on his way to the event. Of how the headlining speaker before him talked for two hours, and he talked for two minutes, and we remember the shorter speech.

All across our country, from the Nation's Capital to the smallest towns, there are military service memorials. Some are simple, some are elaborate, all are necessary. They are there, as Lincoln told us, to be dedicated for those that gave their lives so that a nation may live.

And, what he said then is true today: it is altogether fitting and proper that we should do this.

But, in a larger sense, we can not dedicate — we can not consecrate — we can not hallow — this ground. The brave men, living and dead, who struggled here, have consecrated it, far above our poor power to add or detract.

At the same time, these monuments, these memorials, will never be enough. They will never come close to honoring even a fraction of the sacrifice that was made. That does not mean they are futile...it means that the sacrifice was so great, that no amount of honor could do it justice. No wall of names we can erect, no symbols, no statue can fully describe the measure of sacrifice, the pain of loss, and the pride of service that these monuments represent.

This doesn't mean we shouldn't visit them, touch them. That we shouldn't seek them out. If we stop, then their sacrifice will fade, the importance of their deeds will go unknown. Regardless of how we feel about the wars they fought, their sacrifice should be honored and remembered. I hate war, too[39]...but love and respect those who answered the call. War is an extension of politics by other means, as Von Clauswitz said; but I also recognize that this extension comes with a heavy price, and that price must be paid. Those memorialized in monuments across our nation are the ones who paid the price so we didn't have to.

The world will little note, nor long remember what we say here, but it can never forget what they did here. It is for us the living, rather, to be dedicated here to the unfinished work which they who fought here have thus far so nobly advanced

At the time of the Gettysburg Address, America was in it's youth. The knots and tangles of the Constitution that could not be massaged or removed through politics had exploded into violence. Everyone who died on the Gettysburg battlefield was a countryman of Abraham Lincoln, and I can imagine the enormous pain and sadness he must have felt. The work that was being done to keep this country together was equal in measure to the work that was done to tear this country apart, each side struggling to seize the rights and respect that they felt they deserved. Those who fell in this struggle did not complete this work, only advanced it so that others may complete it.

These memorials remind us: do not go down that path again. Whatever knots, whatever tangles, whatever rights and respect and positions and ideology that we feel must be thrust upon each other can and should be resolved without the need for future memorials. It is to us to continue this work, today, in their name. In their honor. With respect and appreciation for their sacrifice. We are unable to do that if we do not take the time to reflect on their sacrifice.

It is rather for us to be here dedicated to the great task remaining before us — that from these honored dead we take increased devotion to that cause for which they gave the last full measure of devotion

We do have a great task before us: living. Building a better future for each other, for our children. For the children of the fallen, and the children of those who might have fallen but returned. The causes that each service member sacrificed for was so numerous, that any good thing we dedicate ourselves to is one to be honored. They sacrificed for each other. They sacrificed for their families. For their country. Many veterans, including myself, found it an honor to serve so that others would not have to. I served so that my children did not have to. If they choose to do so, then so be it; I will honor them, and pray that they do not pay the ultimate sacrifice.

On this Memorial Day, consider the gravity of the sacrifice of those who fought and died to preserve our way of life. It's not perfect, it's not pretty, but it's ours, and we love it. Like a child slowly realizing the effort that their parents made to provide shelter, food, and a good life, we should take the time to realize the cost of what we have. Money…and Freedom…doesn't grow on trees.

Take some time to visit a local memorial. Trace the names on the walls with your finger. Reflect on the immensity of the cost, not just for that name, but for everyone associated with that name: family, friends, and now you. Standing silently, considering the cost.

Remember.

If the War is Done, Is the Work Done?

On an episode of the Head Space and Timing Podcast, I had a conversation with Sebastian Junger[40]. We were talking about the difficulty of transitioning from the military to post-military life. One of his points: we're currently at point of affluence in our society that allows our nation to provide for those who served in the military without requiring anything more of them.

While we were talking, he referenced the painting, *The Veteran in a New Field*, by Winslow Homer[41]. Homer is considered one of the most significant American artists of the 19th Century; one of his best known paintings is of American Gothic, the picture of the stoic farmer and his wife. Many of the paintings that Homer produced during the Civil War were things that he witnessed when he was sent to the battlefront. This painting was completed in 1865, after the conclusion of the war and the assassination of President Lincoln.

The painting is of a former Union soldier cutting hay in a field. You can see that he's a veteran from the gear that has been set aside in the bottom right corner of the painting; the work of the soldier is complete and the work of the farmer has resumed. While there is certainly are different interpretations of Homer's intent[42] when he created the work, there is also relevance to veterans today.

One Job is Done

To me, one of the most striking concepts about the painting is that this particular veteran's war is over. Even though wars may not end, as evidenced by our current conflicts of almost twenty years, the personal war of some of the service members who fought in them has ended. A war is a terrible thing to pass on to another generation, but our time is complete. That job is done. This veteran, for he is no longer a soldier, has completed their task.

Like the veteran of 150 years ago, the veterans of today have completed their portion of the work. Whether it was large or small, long or short, the service they gave is at an end. For many service members, this is frustrating; they wish they could still be doing the job. Or regret that they stopped doing it. Many feel guilty because they served their time, completed their job, and saw others continuing on. There is a contrast between how we do things today to how it was done in the Civil War and the World Wars; there, the service member stayed until the fighting was done. Today, perhaps another byproduct of the affluence of our society, there is no longer a need for one individual to remain until the war is won.

Another Job Has Begun

Like today, the veteran is relatively young. For those senior leaders who serve thirty years, they may be approaching fifty by the time they leave the service. In a world where the average life expectancy in the U.S. is 78 years old, that's not even to the seventh inning stretch. For many men and women, there is an entire other lifetime ahead of them after leaving the military. Having served twenty-two years in the Army, I know that I will be a veteran twice as long as I was a Soldier, God willing. For many younger veterans, the time spent as a veteran may be as much as four or five times longer than the time they served.

That's a really long time to not do anything. Between my military retirement and VA disability…because nobody gets out of a demolition derby without a few dents and scratches[43]…I really wouldn't have to work. Then there's the challenge of no longer being able to do what we used to do…construction work is a young man's game, unless you're in a supervisory position. Figuring out the next job that we can do, and we want to do, is critical.

The Veteran's Usefulness Does Not End With the War

The other thing that the painting tells me is that there is still work to be done. When a service member leaves the military, they still need to do things; things still need to be done. It may take reinventing ourselves; we may not be able to continue to be laborers, but we can pick up new skills. We can be math teachers. Or authors. Or…if you're interested in psychology…mental health professionals. I'll say here something that was said to me twelve years ago to put me on the path I am now: there are simply not enough combat veterans in the mental health field.

It is truly possible for an old dog to learn new tricks…unless the old dog thinks it's not possible. Winslow Homer's Veteran in a New Field didn't have to reinvent himself; he was likely a farmer before the war, and returned to being a farmer after the war. He could have, though. Just like we can.

Just because the war is over, that doesn't mean the work is over.

Mental Health Professionals Must Be Included in Conversations about Veteran Mental Health

It's been a couple of years since I've been banging the gong about veteran mental health. It's not uncommon for me to bring it up in the first few minutes of a conversation; it doesn't take long to realize where I stand on the issue. "Veteran Mental Health" is quicker to come out of my mouth than a business card out of my hand. The challenge, though, is that I still get the feeling that the message may not be penetrating.

Here is my message, as plainly as I can deliver it: Veteran Mental Health goes beyond just PTSD and TBI[44]. It is a foundational aspect of success in post-military life. Organizations working with veterans should take it into consideration when talking to veterans about their transition plans. Instead, we treat it as an afterthought, something considered when everything else doesn't work.

In my opinion, one of the problems is a lack of understanding of the complexities of veteran mental health. Service members, veterans, and their families don't understand it clearly. Those who haven't served only understand it from how it's presented. Those who do understand and are trying to serve find themselves between a rock and a hard place. The rock: the seemingly endless narrative of the broken veteran. The hard place: the very real needs of the largest cohort of combat veterans since WWII. The problem isn't going anywhere; a recent survey[45] conducted by the Cohen Veterans Network found that the barriers and stigma around veteran mental health aren't going anywhere.

A solution: we need to start looking at Veteran Mental Health as foundational rather than an afterthought. Yes, it includes the bad…PTSD, TBI, and much, much more. It also includes the great: posttraumatic growth[46], Resilience[47], and the very real probability that this generation of Veterans is going to shape this century like the WWII generation shaped the last one. Often, I think, when I talk about "veteran mental health" the message is "all veterans should be in therapy" which translates into "all veterans are crazy." The first is not a fair interpretation of the message, and the second is certainly not true. One of the ways to help change that is to bring mental health professionals into the conversation.

Mental Health Professionals Must Be Included In Transition Conversations

If a community is having a conversation about veteran mental health, and those who have a level of expertise in that field aren't talking…or, if they are talking but few are listening…then a critical piece of the conversation is missing. Often, I have had conversations with people in the community about mental health, often people in charge of making decisions about it, and there are no mental health professionals in the room.

In my conversation with Dr. Heather Kelly on the Head Space and Timing Podcast, she recounted a similar story. In a large meeting of congressional staffers in Washington D.C. that were there to discuss posttrauamtic stress disorder, she quickly realized that not only was she the only person in the room with experience in treating PTSD, but was one of the few in the room that had any connection to the military. When we talk about veteran mental health, and mental health professionals aren't in the room, then any solution that is developed is an imperfect one.

Mental Health Professionals Have A Responsibility to Join the Discussion

At the same time as the community bringing in mental health professionals, mental health professionals can and should engage in the community. One of the things that I have seen over the past several years is that mental health professionals, me included, have a tendency to sit back and wait for veterans to come to us. We know that what we have to offer works. We also forget that we need to convince others that it works. Often, we don't know how to sell it; the industry as a whole doesn't have very good PR.

Another thing I've noticed about the industry: we're not comfortable talking about what we do. Much of that has to do with confidentiality and the sensitive nature of our subject. It also has to do with the fact that it is an industry of introverts. That comes from personal observation, and not research, but the fact is that many of us, again me included, are much more comfortable having deep, meaningful conversations with individuals and small groups. We are much less comfortable engaging in the small talk and networking that is necessary to advocate for a message.

Get Advice from Proven Experts, Not Professed Experts

When I talk about a level of expertise, I'm talking about those who have studied it, and continue to do so. Those who have been able to prove, through testing and accreditation, that they understand mental health. Very often, I see and hear others who talk about mental health and wellness without any specific credentials or degree. It is often those who have experienced a life stressor and have come through the other side of it; that's great, and important. Peer support is a critical component of mental health, in my opinion.

It's when someone uses their experience but not expertise as a badge of honor that things go a little off. There are those that are actually proud of the fact that they're a mental health professional. That it gives them a greater legitimacy than those who haven't. Psychology is a strange thing; so many people are interested in it, and so many people feel as though they are experts in it because they read a book or watched a movie. Other situations aren't like that; if I have a broken leg, I don't go to my cousin for advice about what to do; I go to the emergency room. We tend to do the opposite of that when it comes to mental health and wellness.

So there you have it. To understand the complexities of veteran mental health, as someone who studies it and practices it. Include mental health professionals in the conversation. Involve yourself in the conversation, If you're a mental health professional. As for me, I'm going to continue to do my part to change the way we think about Veteran Mental Health. It ain't pretty, and it ain't perfect, but it has to be done.

As Our Veterans Go, So Goes Our Nation

"The willingness with which our young people are likely to serve in any war, no matter how justified, shall be directly proportional to how they perceive veterans of earlier wars were treated and appreciated by our nation."

This quote, often attributed to George Washington, is used when we hear of mistreatment of veterans and military service members. There is a lot of fist-shaking when we hear of another failure in the treatment of our veterans. There is an uproar when the unemployment numbers are up, or when we consider veteran suicides. That's not particularly what this article is about, though, an attempt to shame the country into behaving better. That tactic didn't work particularly well on me growing up. It probably doesn't work that well on you either. It certainly doesn't seem to make an impact on our communities.

The use of the quote to influence the nation to treat veterans better, or address their needs in a different way, does not reduce the truth in the quote. When we hear it, and roll our eyes, and think, "here we go again, another veteran is wanting special treatment," we're not hearing the words of the quote itself. There is truth in it: how a nation's veterans are treated has a significant impact on whether or not people think serving in the military is an important and valuable endeavor.

In the five years I've been working in veteran support and advocacy, I've heard one question often enough to make it stand out. "What's so special about veterans?" After getting beyond the initial response to the question…often a visceral one…I dig a little deeper, and try to explain. "There's nothing particularly special about *this* veteran," I say, "but veterans as a whole? There is *absolutely* a uniqueness about them."

The deeper question, beyond just the "why should we care," is actually a good one. Why *should* we focus on veterans exclusively when we're trying to solve problems? The national suicide rate is through the roof; it's not just veterans taking their own lives. The problem of homelessness and equitable housing opportunities aren't just a veteran problem. Education, employment, stigma against seeking mental health treatment; all of these are national problems, not veteran problems.

So why give the attention to just veterans? Why pour money into programs that will reduce *veteran* suicide rather than suicide as a whole? That will impact *veteran* employment, rather than a comprehensive employment solution? Doesn't this just widen the "gap" that us veterans are always talking about?

Maybe yes. And, at the same time, no…because the solutions that we apply to the veteran population can then be applied on a larger scale to the nation as a whole.

Veterans are a Diverse and Representative Sample of the Nation

Much of this analysis comes from the 2015 Demographics report[48] from released by the Department of Defense. When I say that our military is representative of our nation, I don't mean that everyone in the nation is represented in the military at the same level that they are in our society. The military continues to be overwhelmingly male. Males make up 84.5% of the total active duty military force in 2015. The number of females in the military in 2015, however, was over 200,000. That's not an insignificant number. The ethnic diversity of the military is comparable to the ethnic diversity of the nation as well. 40% of the total military force consists of racial or ethnic minorities; in 2015, 44% of the nation as a whole were racial and ethnic minorities, according to a Pew Research analysis[49] of the report's data.

All that says this: veterans represent the country. They represent a small portion of our country, but they come from every state in the nation. They come from all walks of life; some are more represented than others, just like some geographic areas[50] are more represented. As any veteran will tell you, however, you quickly adapt to military culture on top of anything else when you enter into the military. The diversity of the military, and by extension the diversity of the veteran population, means that what works with veterans could work with everyone else.

Solutions Developed for Veterans Can Be Applied to All Populations

When you want to solve a big problem, you have to start small. When products come to market, they are tested in small, diverse areas. Research studies rely on representative samples, and if the veteran population is a microcosm of the nation, then they are an identifiable population that can be considered representative. A colleague, reflecting on the "problem" of veterans in higher education, identified that it's not a question of adapting to veteran students; it's about adapting to nontraditional students. Over 75% of college students did not start their college experience directly after high school, according to a recent article[51] discussing the need for colleges and universities to consider nontraditional students in the development of their programs. What can be more nontraditional than a 28 year old (or 42 year old) veteran working their way from an Associate's Degree to a Master's Degree?

The same can be applied to employment. The veteran population, and by extension, their families, are highly mobile and diverse. Not only is unemployment a challenge for the veteran population, so is underemployment: the situation where an individual's job is not taking full advantage of their skills and experience. This, too, can be applied to the nation as a whole: college graduates who start out underemployed are likely to stay that way, according to another recent research study[52]. Of course, any job is better than no job, and getting a job is easier if you have a job, but chronic and sustained underemployment is a challenge in the veteran community. If we can solve it there, we can apply that solution to other areas.

And so on, and so on. Mental health stigma? Absolutely, going strong in the veteran community. If we can get veterans to change the way they think and talk about mental health, then we can apply those same techniques to the nation as a whole. Veteran homelessness solutions can be applied to homeless populations overall.

Strengthening Our Former Service Members Strengthens Our Nation

And yes, finally, back to the quote above. Taking care of our nation's veterans is very much a national security issue. I've written about this before, but in his book *Tribe*[53], Sebastian Junger writes, "Humans are so strongly wired to help one another – and enjoy such enormous social benefits from doing so – that people regularly risk their lives for complete strangers." What happens when that strong wiring conflicts with an equal desire for self-preservation? When there are no longer social benefits, but actual social detriments to helping others? We start to get a nation of the served rather than the serving.

And that, my friends, is a truly frightening future.

Acknowledgements

When it comes to support, there are often too many important people to mention and not enough space to do so. As always, there is my family, who have stood by me during the long days and busy times. As my wife has said, she thought she would see more of me after I retired from the Army, not less of me; but she knows how important the work is to me, and how important it is to others. And to my two children, who were just entering elementary school when we came to Colorado; thirteen years later, they're entering college. Their support and encouragement has been outstanding.

My colleagues at the Family Care Center, as well as the various organizations I'm affiliated with in Colorado Springs, have been nothing but supportive and encouraging as well. Their dedication to the mental health and wellness of service members, veterans, and their families is unquestionable, and greatly appreciated.

While this book was being written, I had the honor and privilege to be selected to attend the inaugural cohort of the George W. Bush Institute Stand-To Veteran Leadership Program. I met thirty-two amazing people, and if I were to name them by name, we would be here for a heck of a long time. If I were to describe the impact that they have had on me, we would be here even longer.

On the day that I write this, I got the message that we all dread: a fellow veteran was in crisis, and we might be too late. The frustration, even the anger, became paralyzing: not again. But the unfailing and never-quit attitude of a huge supporter of the work that I do finally got help to the place where it was needed. In this, she consistently shows the dedication to helping her fellow veterans; if it weren't for her support, I might have given up on Head Space and Timing a long time ago. She puts a metaphorical boot in my butt when it's needed, and makes sure I change my socks and drink water, just like any good medic will. Thanks, Doc Johnson.

I have a lot of loyal followers of the blog and podcast, and am consistently amazed when I hear of someone who shares something that I wrote or produced. The messages I get of how these words made an impact are worth everything; keep 'em coming.

If you are serving, have served, are a military family member, remember…you are not alone. Ever.

Notes

Introduction

1. Market segmentation is a term that describes grouping people into categories with common needs and desires
   https://www.investopedia.com/terms/m/marketsegmentation.asp
2. See Moral Injury: The Impact of Combat on Veteran's Individual Morality
   http://veteranmentalhealth.com/moral-injury-the-impact-of-combat-on-veterans-individual-morality/
3. Veterans and Maslow's Hierarchy of Needs, Head Space and Timing Blog, 12 Aug 2016
   http://veteranmentalhealth.com/veterans-and-maslows-hierarchy-of-needs/
4. Joshua Kreimeyer, Army combat veteran and licensed marriage and family therapist, talks about family and mental health on episode 35 of the Head Space and Timing podcast http://veteranmentalhealth.com/hst035/
5. Army combat veteran Meaghan Mobbs, "Veteran Mental Health is Not One Size Fits All" 11 June 2018 https://www.psychologytoday.com/us/blog/the-debrief/201806/veteran-mental-health-is-not-one-size-fits-all

Part I Honest Discussions about Veteran Suicide

1. Nock, M. K., Stein, M. B., Heeringa, S. G., Ursano, R. J., Colpe, L. J., Fullerton, C. S., ... & Zaslavsky, A. M. (2014). Prevalence and correlates of suicidal behavior among soldiers: results from the Army Study to Assess Risk and Resilience in Servicemembers (Army STARRS). *JAMA psychiatry, 71*(5), 514-522.
2. A collection of Head Space and Timing suicide prevention articles can be found at http://veteranmentalhealth.com/national-suicide-prevention-week-2017/
3. Army veteran Tony Williams, guest on episode 17 of the Head Space and Timing Podcast, October 24th, 2017 http://veteranmentalhealth.com/hst107-2/
4. Dr. Stacey Freedenthal is a psychologist, suicidologist, associate professor at the University of Denver Graduate School of Social Work and founder of www.speakingofsuicide.com
5. Episode 43 of the Head Space and Timing Podcast, A Serious Look at Suicide http://veteranmentalhealth.com/043/
6. 10 Things Not to Say to a Suicidal Person by Stacey Freedenthal https://www.speakingofsuicide.com/2015/03/03/what-not-to-say/

Part 2: Who We Were

1. For Veterans, Remembering Where You Came From is Key to a Successful Transition, April 6, 2016 http://veteranmentalhealth.com/for-veterans-remembering-where-you-came-from-is-key-to-a-successful-transition/
2. The full Soldiers Creed can be found at https://www.army.mil/values/soldiers.html
3. Combat: It Was the Best of Times, It Was the Worst of Times, Head Space and Timing Blog, 15 February 2016 http://veteranmentalhealth.com/combat-it-was-the-best-of-times-it-was-the-worst-of-times/
4. The Combat Veteran Paradox: An Infographic, Head Space and Timing Blog, 26 January 2017 http://veteranmentalhealth.com/the-combat-veteran-paradox-an-infographic/
5. The Combat Addiction Paradox, Head Space and Timing Blog, 21 February 2017 http://veteranmentalhealth.com/the-combat-addiciton-paradox/
6. The First Battle of Fallujah occurred from April 4- May 1, 2004, and the Second Battle of Fallujah occurred from November 7-December 23, 2004. Insurgent forces battling Iraqi National Army captured the city in 2014

7. What the Teen who Vandalized the Medal of Honor Memorial Deserves, by Brian Thompson, 27 July 2018 https://www.nydailynews.com/opinion/ny-oped-what-the-vandal-deserves-20180727-story.html
8. Metallica at War: Combat and Confusion, Head Space and Timing Podcast, 24 November 2016 http://veteranmentalhealth.com/metallica-at-war-combat-and-confusion/
9. An Open Letter to America, from One of Your Veterans, Head Space and Timing Blog, 9 June 2016 http://veteranmentalhealth.com/an-open-letter-to-america-from-one-of-your-veterans/
10. Frankl, V. E. (1985). *Man's search for meaning*. Simon and Schuster.
11. A collection of articles and podcasts on the Head Space and Timing blog, http://veteranmentalhealth.com/tag/moral-injury/
12. Rick Rescorla: Leadership Done Right, an episode of the Head Space and Timing Podcast with Jeff Adamec, https://share.transistor.fm/s/2655575d
13. Episode 83 of the Head Space and Timing Podcast, Building Mental Fitness with Dr. Kate Hendricks Thomas http://veteranmentalhealth.com/hst083/
14. I Kept My Abusive Marriage a Secret Because Marines Are Supposed to be Tough, Task and Purpose, 1 November 2016 https://taskandpurpose.com/kept-abusive-marriage-secret-marines-supposed-tough
15. Fight or Flight: The Veterans at War with PTSD, The Guardian, 8 November 2018 https://www.youtube.com/watch?v=ps4hKECvNMg&feature=youtu.be
16. Joseph Heller, 1 May 1923-12 December 1999, was an author and veteran of the Army Air Corps Heller, J. (1961). *Catch-22: a novel*. Simon and Schuster.
17. As Good as I Once Was, Toby Keith, 16 June 2009
18. Stop Comparing Your Behind-The-Scenes with Everyone's Highlight Reel, Carrie Kerpen, Forbes https://www.forbes.com/sites/carriekerpen/2017/07/29/stop-comparing-your-behind-the-scenes-with-everyones-highlight-reel/#731505d13a07
19. Helping Veterans Trapped by Their Own Experiences: Learned Helplessness and Veteran Mental Health, Head Space and Timing Blog, 6 Aug 2016 http://veteranmentalhealth.com/helping-veterans-trapped-by-their-own-experiences-learned-helplessness-and-veteran-mental-health/
20. The Dangerous Trap of the Comparison Game, Head Space and Timing Blog, 20 September 2018 http://veteranmentalhealth.com/comparison-game/
21. The Story of Steve Jobs, Xerox, and Who Really Invented the Personal Computer, Newsweek.com https://www.newsweek.com/silicon-valley-apple-steve-jobs-xerox-437972
22. Maybe the Toughest Man Alive, Defense.gov https://www.defense.gov/explore/story/Article/1707737/maybe-the-toughest-man-alive/
23. Is Anxiety about Shadows and Legends Getting In Your Way, Head Space and Timing Blog, 14 March 2017 http://veteranmentalhealth.com/is-anxiety-about-shadows-and-legends-getting-in-your-way/
24. The Dangers and Benefits of Living a Private Life in Public, Head Space and Timing Blog, 8 June 2017 http://veteranmentalhealth.com/dangers-and-benefits-of-living-a-private-life-in-public/
25. Heroes & Monsters: War's moral injury, Sebastian J. Bae, Foreign Policy.com https://foreignpolicy.com/2015/02/27/heroes-monsters-wars-moral-injury/
26. Pol, E., Di Masso, A., Castrechini, A., Bonet, M., & Vidal, T. (2006). Psychological parameters to understand and manage the NIMBY effect. *Revue Européenne de Psychologie Appliquée/European Review of Applied Psychology, 56*(1), 43-51.
27. V is for Veteran, Not Villain, Victim, or Vindicator, Head Space and Timing Blog, 19 July 2017 http://veteranmentalhealth.com/v-is-for-veteran-not-villain-victim-or-vindicator/

28. On Being 'One of Those Weird Veterans' in the Workplace, guest post on the Head Space and Timing blog by veteran Garrett Wilkerson, 14 September 2016 http://veteranmentalhealth.com/on-being-one-of-those-weird-veterans-in-the-workplace/

29. Rudstam, H., Strobel Gower, W., & Cook, L. (2012). Beyond yellow ribbons: Are employers prepared to hire, accommodate and retain returning veterans with disabilities?. *Journal of Vocational Rehabilitation*, *36*(2), 87-95.

30. The Veteran Divide, Head Space and Timing Blog, 22 September 2016 http://veteranmentalhealth.com/the-veteran-divide/

31. Foley, P. S. (2014). The metaphors they carry: Exploring how veterans use metaphor to describe experiences of PTSD and the implications for social work practice.

32. Blackwell-Starnes, K. (2018). At ease: Developing veterans' sense of belonging in the college classroom. *Journal of Veterans Studies*, *3*(1), 18-36.

33. Seeking Serendipity: Paying Attention to Beneficial Chance, Head Space and Timing Blog, 18 October 2018 http://veteranmentalhealth.com/serendipity/

34. For Veterans, Our Capacity for Stress is Greater Than We Know, Head Space and Timing Blog, 2 June 2016 http://veteranmentalhealth.com/for-veterans-our-capacity-for-stress-is-greater-than-we-know/

35. Kubrick, S., Herr, M., & Hasford, G. (2001). *Full metal jacket*. Burbank, CA: Warner Home Video.

36. Episode 3 of Band of Brothers, a special podcast series on the Change Your POV Podcast http://changeyourpov.libsyn.com/podcast/bob003-day-of-days

37. Tull, M. T., & Roemer, L. (2003). Alternative explanations of emotional numbing of posttraumatic stress disorder: An examination of hyperarousal and experiential avoidance. *Journal of Psychopathology and Behavioral Assessment*, *25*(3), 147-154.

38. The Violence of Action Paradox: Emotional Contradiction of Veterans, Head Space and Timing Blog, 19 February 2016 http://veteranmentalhealth.com/the-violence-of-action-paradox-emotional-contradiction-of-veterans/

39. Kevin Sullivan is the Senior Advisor, External Affairs, George W. Bush Presidential Center https://www.bushcenter.org/people/kevin-sullivan.html

40. Episode 82 of the Head Space and Timing Podcast, Prolonged Exposure Research with Dr. Carmen McLean http://veteranmentalhealth.com/hst082/

41. Belova, M. A., Paton, J. J., Morrison, S. E., & Salzman, C. D. (2007). Expectation modulates neural responses to pleasant and aversive stimuli in primate amygdala. *Neuron*, *55*(6), 970-984.

42. Cognitive Impairment and the Neurological Basis for PTSD, Head Space and Timing Blog, 20 July 2017 http://veteranmentalhealth.com/cognitive-impairment-and-the-neurological-basis-for-ptsd/

43. The Things They Carry Now, Head Space and Timing Blog, 17 February 2016 http://veteranmentalhealth.com/the-things-they-carry-now/

44. Sullivan, M. J., Rodgers, W. M., & Kirsch, I. (2001). Catastrophizing, depression and expectancies for pain and emotional distress. *Pain*, *91*(1-2), 147-154.

45. Ward, A., Lyubomirsky, S., Sousa, L., & Nolen-Hoeksema, S. (2003). Can't quite commit: Rumination and uncertainty. *Personality and social psychology bulletin*, *29*(1), 96-107.

46. Episode 99 of the Head Space and Timing Podcast: The Alchemy of Combat with Dr. Larry Decker http://veteranmentalhealth.com/hst099/

47. Teeters, J. B., Lancaster, C. L., Brown, D. G., & Back, S. E. (2017). Substance use disorders in military veterans: prevalence and treatment challenges. *Substance abuse and rehabilitation*, *8*, 69.

48. Monson, C. M., Taft, C. T., & Fredman, S. J. (2009). Military-related PTSD and intimate relationships: From description to theory-driven research and intervention development. *Clinical psychology review*, *29*(8), 707-714.

49. Maladaptive behavior examples https://www.psysci.co/maladaptive-behavior-examples/
50. 10 Thinking Errors that will Crush Your Mental Strength, Amy Morin on Psychology Today, 24 January 2015 https://www.psychologytoday.com/us/blog/what-mentally-strong-people-dont-do/201501/10-thinking-errors-will-crush-your-mental-strength
51. Taylor, D. J., Lichstein, K. L., Durrence, H. H., Reidel, B. W., & Bush, A. J. (2005). Epidemiology of insomnia, depression, and anxiety. *Sleep*, *28*(11), 1457-1464.
52. Hughes, J. M., Ulmer, C. S., Gierisch, J. M., Hastings, S. N., & Howard, M. O. (2018). Insomnia in United States military veterans: An integrated theoretical model. *Clinical psychology review*, *59*, 118-125.
53. Marmar, C. R., Schlenger, W., Henn-Haase, C., Qian, M., Purchia, E., Li, M., ... & Karstoft, K. I. (2015). Course of posttraumatic stress disorder 40 years after the Vietnam War: Findings from the National Vietnam Veterans Longitudinal Study. *JAMA psychiatry*, *72*(9), 875-881.
54. Perlis, M. L., Giles, D. E., Mendelson, W. B., Bootzin, R. R., & Wyatt, J. K. (1997). Psychophysiological insomnia: the behavioural model and a neurocognitive perspective. *Journal of sleep research*, *6*(3), 179-188.
55. Riemann, D., Spiegelhalder, K., Feige, B., Voderholzer, U., Berger, M., Perlis, M., & Nissen, C. (2010). The hyperarousal model of insomnia: a review of the concept and its evidence. *Sleep medicine reviews*, *14*(1), 19-31.
56. Nappi, C. M., Drummond, S. P., & Hall, J. M. (2012). Treating nightmares and insomnia in posttraumatic stress disorder: a review of current evidence. *Neuropharmacology*, *62*(2), 576-585.
57. Taylor, D. J., & Pruiksma, K. E. (2014). Cognitive and behavioural therapy for insomnia (CBT-I) in psychiatric populations: a systematic review. *International review of psychiatry*, *26*(2), 205-213.
58. Krakow, B., & Zadra, A. (2006). Clinical management of chronic nightmares: imagery rehearsal therapy. *Behavioral sleep medicine*, *4*(1), 45-70.
59. Ruff, R. L. (2009). Improving sleep: initial headache treatment in OIF/OEF veterans with blast-induced mild traumatic brain injury. *Journal of rehabilitation research and development*, *46*(9), 1071.
60. Learning Self-Care in Your Post-Military Life, Head Space and Timing Blog, 28 September 2017 http://veteranmentalhealth.com/learning-self-care-post-military-life/
61. Linehan, M. (2014). *DBT Skills training manual*. Guilford Publications.
62. Comprehensive Veteran Mental Health Addresses All Aspects, Head Space and Timing Blog, 22 February 2018 http://veteranmentalhealth.com/comprehensive/
63. The Veteran Divide, Head Space and Timing Blog, 22 September 2016 http://veteranmentalhealth.com/the-veteran-divide/
64. A Message From A Veteran To Veterans: You Have the Potential to Change The World, Head Space and Timing Blog, 11 November 2016 http://veteranmentalhealth.com/a-message-from-a-veteran-to-veterans-you-have-the-potential-to-change-the-world/
65. Modern Veterans Do Not Need to Become Another 'Lost Generation', Christopher Kuhn, 14 October 2016, Task & Purpose, https://taskandpurpose.com/modern-veterans-do-not-need-to-become-another-lost-generation
66. Welcome to the First Cross-Generational War, Head Space and Timing Blog, 2 May 2017 http://veteranmentalhealth.com/welcome-to-the-first-cross-generational-war/

Part 3: Who We Are

1. The Arkham Sessions, hosted by Dr. Andrea Letamendi and Brian Ward, is a weekly podcast dedicated to the psychological analysis of Batman: The Animated Series. http://www.underthemaskonline.com/the-arkham-sessions/

2.   Veterans, Do You Long For Days Gone By? Head Space and Timing Blog, 4 April 2017 http://veteranmentalhealth.com/veterans-do-you-long-for-days-gone-by/

3.   Ajzen, I. (2002). Perceived behavioral control, self-efficacy, locus of control, and the theory of planned behavior 1. *Journal of applied social psychology, 32*(4), 665-683.

4.   Whealin, J. M., Kuhn, E., & Pietrzak, R. H. (2014). Applying behavior change theory to technology promoting veteran mental health care seeking. *Psychological Services, 11*(4), 486.

5.   LeardMann, C. A., Smith, B., & Ryan, M. A. (2010). Do adverse childhood experiences increase the risk of postdeployment posttraumatic stress disorder in US Marines?. *BMC Public Health, 10*(1), 437.

6.   Lilly, M. M., Valdez, C. E., & Graham-Bermann, S. A. (2011). The mediating effect of world assumptions on the relationship between trauma exposure and depression. *Journal of Interpersonal Violence, 26*(12), 2499-2516.

7.   How to Completely Change How You See The World, Head Space and Timing Blog, 10 January 2017 http://veteranmentalhealth.com/how-to-completely-change-how-you-see-the-world/

8.   Mobbs, M. C., & Bonanno, G. A. (2018). Beyond war and PTSD: The crucial role of transition stress in the lives of military veterans. *Clinical psychology review, 59*, 137-144.

9.   About Face is a website produced by the Department of Veterans Affairs to reduce the stigma against PTSD and the psychological impact of trauma https://www.ptsd.va.gov/apps/AboutFace/

10. PsychArmor is a 501(c)(3) nonprofit that offers critical resources to Americans so they can effectively engage with and support military service members, Veterans, and their families. www.psycharmor.org

11. Ribeiro, J. D., & Joiner, T. E. (2009). The interpersonal-psychological theory of suicidal behavior: Current status and future directions. *Journal of clinical psychology, 65*(12), 1291-1299.

12. Cohen, B. E., Gima, K., Bertenthal, D., Kim, S., Marmar, C. R., & Seal, K. H. (2010). Mental health diagnoses and utilization of VA non-mental health medical services among returning Iraq and Afghanistan veterans. *Journal of general internal medicine, 25*(1), 18-24.

13. Phelan, S. M., Griffin, J. M., Hellerstedt, W. L., Sayer, N. A., Jensen, A. C., Burgess, D. J., & Van Ryn, M. (2011). Perceived stigma, strain, and mental health among caregivers of veterans with traumatic brain injury. *Disability and Health Journal, 4*(3), 177-184.

14. Jordan, B. K., Marmar, C. R., Fairbank, J. A., Schlenger, W. E., Kulka, R. A., Hough, R. L., & Weiss, D. S. (1992). Problems in families of male Vietnam veterans with posttraumatic stress disorder. *Journal of consulting and clinical psychology, 60*(6), 916.

15. Stein, J. Y., & Tuval-Mashiach, R. (2015). The social construction of loneliness: An integrative conceptualization. *Journal of Constructivist Psychology, 28*(3), 210-227.

16. The Dangers and Benefits of Living a Private Life in Public, Head Space and Timing Blog, 8 June 2017 http://veteranmentalhealth.com/dangers-and-benefits-of-living-a-private-life-in-public/

17. Avoiding Avoidance Leads to Success in Post-Military Life, Head Space and Timing Blog, 6 December 2018 http://veteranmentalhealth.com/avoidance/

18. Yalom, I. D. (2008). Staring at the sun: Overcoming the terror of death. *The Humanistic Psychologist, 36*(3-4), 283-297.

19. Steiner, A. P., & Redish, A. D. (2014). Behavioral and neurophysiological correlates of regret in rat decision-making on a neuroeconomic task. *Nature neuroscience, 17*(7), 995.

20. Dweck, C. S. (2008). *Mindset: The new psychology of success.* Random House Digital, Inc..

21. Through the Other Side of the Valley of Death: Veterans and Posttraumatic Growth, Head Space and Timing Blog, 9 March 2016 http://veteranmentalhealth.com/through-the-other-side-of-the-valley-of-death-veterans-and-posttraumatic-growth/

22. The Troubles of Our Presidents, by Frederic Austin Ogg, Munsey's Magazine, Volume 43

23. Episode 125 of the Head Space and Timing Podcast, Dr. Jannell MacAulay: Positive Mindset and Mental Fitness http://veteranmentalhealth.com/hst125/

24. Trapped in a Cage of Our Own Construction, Head Space and Timing Blog, 14 February 2017 http://veteranmentalhealth.com/trapped-in-a-cage-of-our-own-construction/

25. Words to Use with Caution: 'But', Pegasus NLP Web site, https://nlp-now.co.uk/be-careful-with-but/

26. 5 Keys to Successful Military Transition from Positive Psychology, Head Space and Timing Blog, 29 July 2017 http://veteranmentalhealth.com/5-keys-to-successful-military-transition-from-positive-psychology/

27. Our Brain's Negative Bias, by Hara Estroff Marano, Psychology Today 20 June 2003, https://www.psychologytoday.com/us/articles/200306/our-brains-negative-bias

28. Garland, E. L., Fredrickson, B., Kring, A. M., Johnson, D. P., Meyer, P. S., & Penn, D. L. (2010). Upward spirals of positive emotions counter downward spirals of negativity: Insights from the broaden-and-build theory and affective neuroscience on the treatment of emotion dysfunctions and deficits in psychopathology. *Clinical psychology review, 30*(7), 849-864.

29. Ibid.

30. Brené Brown is a research professor at the University of Houston and the author of five New York Times Bestsellers, https://brenebrown.com

31. 3 Easiest Resiliency Skills Ever, Military.com, https://www.military.com/spousebuzz/blog/2014/05/3-easiest-resiliency-skills-ever.html

32. The Warrior Awareness and Accountability Log, Head Space and Timing Blog, http://veteranmentalhealth.com/wp-content/uploads/2016/06/THE-WARRIOR-RECOVERY-PLAN.pdf

33. Pandora's Box of the Veteran Mind, Head Space and Timing Blog, http://veteranmentalhealth.com/the-pandoras-box-of-the-veteran-mind/

34. Penn, N. E., Kramer, J., Skinner, J. F., Velasquez, R. J., Yee, B. W., Arellano, L. M., & Williams, J. P. (2000). Health practices and health-care systems among cultural groups. *Handbook of gender, culture, and health*, 105-138.

35. "Slut, Bitch, Dyke" – Joan of Arc and the Modern Military Woman, Mariette Kalinowski, 21 August, 2013 The Daily Beast Web site https://www.thedailybeast.com/slut-bitch-dykejoan-of-arc-and-the-modern-military-woman

36. The Marine Nude-Photo Scandal is Growing and Adding New Victims, James LaPorta and Rory Laverty, 22 March 2017, The Daily Best Web site, https://www.thedailybeast.com/the-marine-nude-photo-scandal-is-growing-and-adding-new-victims

37. Mushtaq, M., Sultana, S., & Imtiaz, I. (2015). The trauma of sexual harassment and its mental health consequences among nurses. *Journal of the College of Physicians and Surgeons Pakistan, 25*(9), 675-679.

38. Reed, G. E. (2004). Toxic leadership. *Military review, 84*(4), 67-71.

39. Helping Veterans Trapped By Their Own Experiences: Learned Helplessness and Veteran Mental Health, Head Space and Timing Blog, 6 August 2016 http://veteranmentalhealth.com/helping-veterans-trapped-by-their-own-experiences-learned-helplessness-and-veteran-mental-health/

40. Taylor, P. J., Gooding, P., Wood, A. M., & Tarrier, N. (2011). The role of defeat and entrapment in depression, anxiety, and suicide. *Psychological bulletin, 137*(3), 391.

41. Ruderman, L., Ehrlich, D. B., Roy, A., Pietrzak, R. H., Harpaz-Rotem, I., & Levy, I. (2016). Posttraumatic stress symptoms and aversion to ambiguous losses in combat veterans. *Depression and anxiety*, *33*(7), 606-613.

42. Google's Veteran Tool – They Got It Wrong! David Lee, 19 September 2018 https://www.linkedin.com/pulse/googles-veteran-tool-got-wrong-david-lee/

43. Rubin, V. L., Burkell, J., & Quan-Haase, A. (2011). Facets of serendipity in everyday chance encounters: a grounded theory approach to blog analysis. *Information Research*, *16*(3).

44. Four Keys to Navigating Obstacles on a River…and in Life, Head Space and Timing Web Site, 11 May 2017 http://veteranmentalhealth.com/four-keys-to-navigating-obstacles-on-a-river-and-in-life/

45. How to Completely Change How You See the World, Head Space and Timing Blog, 10 January 2017 http://veteranmentalhealth.com/how-to-completely-change-how-you-see-the-world/

46. Thankfulness for Transition and Change, Head Space and Timing Blog, 21 November 2018, http://veteranmentalhealth.com/transition/

47. The Violence of Action Paradox: Emotional Contradiction of Veterans, Head Space and Timing Blog, 19 February 2016, http://veteranmentalhealth.com/transition/

48. Linehan, M. (2014). *DBT Skills training manual*. Guilford Publications.

49. Charpentier, C. J., Aylward, J., Roiser, J. P., & Robinson, O. J. (2017). Enhanced risk aversion, but not loss aversion, in unmedicated pathological anxiety. *Biological psychiatry*, *81*(12), 1014-1022.

50. Ibid

51. Episode 27 of the Head Space and Timing Podcast, External Barriers to Veteran Mental Health with Erin Fowler, http://veteranmentalhealth.com/hst027-external-barriers-veteran-mental-health-erin-fowler/

52. Starving at the Feast: A Parable of Military Transition, Head Space and Timing Blog, 27 July 2017 http://veteranmentalhealth.com/starving-at-the-feast-a-parable-of-military-transition/

53. Revisions to the Psychological and Emotional Health Questions on the Standard Form 86, Questionnaire for National Security Positions, 16 November 2016

54. Superhero in the House: Living with Your Service Member's Alter Ego, Head Space and Timing Blog, 9 February 2017 http://veteranmentalhealth.com/superhero-in-the-house-living-with-your-service-members-alter-ego/

55. Kulesza, M., Pedersen, E. R., Corrigan, P. W., & Marshall, G. N. (2015). Help-seeking stigma and mental health treatment seeking among young adult veterans. *Military behavioral health*, *3*(4), 230-239.

56. Episode 47 of the Head Space and Timing Podcast, Why So Serious with Rhiannon Guzelian, http://veteranmentalhealth.com/hst047/

57. Person-Centered Therapy, Good Therapy Web site, https://www.goodtherapy.org/learn-about-therapy/types/person-centered

58. Getting Perspective in Post-Military Life, Head Space and Timing Blog, 21 June 2018 http://veteranmentalhealth.com/perspective/

59. Christopher Lochhead Follow Your Different Podcast, https://podcasts.apple.com/us/podcast/christopher-lochhead-follow-your-different/id1204044507

60. Feldman, M. D., Franks, P., Duberstein, P. R., Vannoy, S., Epstein, R., & Kravitz, R. L. (2007). Let's not talk about it: suicide inquiry in primary care. *The Annals of Family Medicine*, *5*(5), 412-418.

61. Suicide Rates Rising Across the U.S., 7 June 2018, Centers for Disease Control Web site https://www.cdc.gov/media/releases/2018/p0607-suicide-prevention.html

62. Brashers, D. E., Goldsmith, D. J., & Hsieh, E. (2002). Information seeking and avoiding in health contexts. *Human Communication Research*, *28*(2), 258-271.

63. Awareness is the Key to Recovery for Veterans, Head Space and Timing Blog, 5 February 2016 http://veteranmentalhealth.com/awareness-is-the-key-to-recovery-for-veterans/

64. Sex, Religion & Politics: Why You Should Never Discuss Them At Work, Glassdor.com Web site, 29 August 2012 https://www.glassdoor.com/blog/sex-religion-politics-stay-discussing-office/

65. Dailey, R. M., & Palomares, N. A. (2004). Strategic topic avoidance: An investigation of topic avoidance frequency, strategies used, and relational correlates. *Communication Monographs, 71*(4), 471-496.

66. For Veterans, a Rest Plan is Critical to Future Success, Head Space and Timing Blog, 12 May 2016, http://veteranmentalhealth.com/for-veterans-a-rest-plan-is-critical-to-future-success-2/

Part 4: Who We Will Become

1. Meaning, Purpose, and Veteran Mental Health, Head Space and Timing Blog, 18 August 2016, http://veteranmentalhealth.com/meaning-purpose-and-veteran-mental-health/

2. Bluck, S., & Liao, H. W. (2013). I was therefore I am: Creating self-continuity through remembering our personal past. *The International Journal of Reminiscence and Life Review, 1*(1), 7-12.

3. Ibid

4. Veterans and the Rubber Bullets of Our Thoughts, Head Space and Timing Blog, 21 July 2016, http://veteranmentalhealth.com/veterans-and-the-rubber-bullets-of-our-thoughts/

5. Steeg, S., Haigh, M., Webb, R. T., Kapur, N., Awenat, Y., Gooding, P., ... & Cooper, J. (2016). The exacerbating influence of hopelessness on other known risk factors for repeat self-harm and suicide. *Journal of affective disorders, 190*, 522-528.

6. Avoiding Pain Because of the Fear of Suffering, Head Space and Timing Blog, 3 January 2019 http://veteranmentalhealth.com/suffering/

7. I Never Saw Combat as a Marine in Afghanistan, and I'm Okay With That, Geoff Dempsey, 20 October 2016, https://taskandpurpose.com/i-never-knew-what-real-boredom-was-until-i-deployed-to-afghanistan

8. Go With the Flow, John Gierland, WIRED Web site, 1 September 96, https://www.wired.com/1996/09/czik/

9. Nakamura, J., & Csikszentmihalyi, M. (2002). The concept of flow. Handbook of positive psychology, 89-105

10. Meaning, Purpose, and Veteran Mental Health, Head Space and Timing Blog, 18 August 2016, http://veteranmentalhealth.com/meaning-purpose-and-veteran-mental-health/

11. The Beginner's Guide to Shamming, Nine Line Web site, 11 February 2017 http://ninelinenews.com/articles-2/military/the-beginners-guide-to-shamming/

12. How to Completely Change How You See The World, Head Space and Timing Blog, 10 January 2017 http://veteranmentalhealth.com/how-to-completely-change-how-you-see-the-world/

13. Why are the Younger Veterans Avoiding the Veteran Service Organizations? Les Davis, U.S. Veterans Magazine Web site, https://www.usveteransmagazine.com/2017/02/why-are-the-younger-veterans-avoiding-the-veteran-service-organizations/

14. The Veterans of Foreign Wars of the United States is a nonprofit veterans service organization comprised of eligible veterans and military service members from the active, guard and reserve forces. www.vfw.org

15. The American Legion was chartered and incorporated by Congress in 1919 as a patriotic veterans organization devoted to mutual helpfulness. It is the nation's largest

wartime veterans service organization, committed to mentoring youth and sponsorship of wholesome programs in our communities, advocating patriotism and honor, promoting strong national security, and continued devotion to our fellow servicemembers and veterans.

16. Vietnam Veterans of America's goals are to promote and support the full range of issues important to Vietnam veterans, to create a new identity for this generation of veterans, and to change public perception of Vietnam veterans. www.vva.org

17. Paralyzed Veterans of America is a Nonprofit that has a mission to change lives and build brighter futures for seriously injured veterans www.pva.org

18. Team Rubicon is a 501(c)3 nonprofit that utilizes the skills and experiences of military veterans with first responders to rapidly deploy emergency response teams. https://teamrubiconusa.org/

19. The Travis Manion Foundation is strengthening America's national character. In an increasingly divisive cultural climate, we unite and strengthen communities. Our nation needs role models that inspire and we should look no further than the civic assets already living in our communities. Veterans and families of the fallen are leading the charge, pushing us to be better versions of ourselves and improving our collective character. https://www.travismanion.org/

20. What is Entrepreneurship? Entrepreneurship Definition and Meaning, Nicole Ferreira, Oberlo Web site, 12 May 2019 https://www.oberlo.com/blog/what-is-entrepreneurship

21. The primary mission of High Ground Veterans Advocacy (HGVA) is to improve the lives of servicemembers, veterans and their families through education, charitable action, and scientific research. www.highgroundvets.org

22. The Benefit of Gratitude to Veteran Mental Health, Head Space and Timing Blog, 23 November 2017, http://veteranmentalhealth.com/benefit-gratitude-veteran-mental-health/

23. What does "Dayenu" Mean Today? Joshua Ratner, My Jewish Learning Web site, 1 April 2014, https://www.myjewishlearning.com/rabbis-without-borders/what-does-dayenu-mean-today/

24. Trapped in a Cage of Our Own Construction, Head Space and Timing Web site, http://veteranmentalhealth.com/trapped-in-a-cage-of-our-own-construction/

25. Service to School's mission is to prepare transitioning military veterans for their next chapter of leadership by helping them gain admission to the best college or graduate school possible. www.service2school.org

26. The Paradox of the Veteran Story, Head Space and Timing Web site, 15 September 2016 http://veteranmentalhealth.com/the-paradox-of-the-veteran-story/

27. Background Knowledge: The Glue that Makes Learning Stick, ReLea Lent, Overcoming Textbook Fatigue, http://veteranmentalhealth.com/the-paradox-of-the-veteran-story/

28. Adams, R. E., Urosevich, T. G., Hoffman, S. N., Kirchner, H. L., Hyacinthe, J. C., Figley, C. R., ... & Boscarino, J. A. (2017). Social support, help-seeking, and mental health outcomes among veterans in non-VA facilities: results from the veterans' health study. *Military behavioral health*, *5*(4), 393-405.

29. National Academies of Sciences, Engineering, and Medicine. (2018). *Evaluation of the department of veterans affairs mental health services*. National Academies Press.

30. Castro, C. A., & Kintzle, S. (2014). Suicides in the military: the post-modern combat veteran and the Hemingway effect. *Current Psychiatry Reports*, *16*(8), 460.

31. Public Perception of Veterans and Veteran Mental Health, Head Space and Timing Blog, 24 January 2019, http://veteranmentalhealth.com/public-perception-of-veterans-and-veteran-mental-health/

32. Veterans on Wall Street (VOWS) is an initiative dedicated to honoring former and current military personnel by facilitating career and business opportunities in the financial services industry. https://veteransonwallstreet.com

33. Episode 99 of the Head Space and Timing Podcast: The Alchemy of Combat with Dr. Larry Decker http://veteranmentalhealth.com/hst099/

34. The Veteran Divide, Head Space and Timing Blog, 22 September 2016 http://veteranmentalhealth.com/the-veteran-divide/

35. Public Perception of Veterans and Veteran Mental Health, Head Space and Timing Blog, 24 January 2019, http://veteranmentalhealth.com/public-perception-of-veterans-and-veteran-mental-health/

36. Finding Motivation, Aaron Skogen, Guest Post on the Head Space and Timing Blog, 14 December 2016, http://veteranmentalhealth.com/finding-motivation/

37. The Path of Painful Truth, Head Space and Timing Blog, 12 October 2017, http://veteranmentalhealth.com/path-painful-truth/

38. The Vietnam Memorial Traveling Wall is a 3/5 scale replica of the Vietnam Memorial in Washington, DC. http://www.travelingwall.us

39. The Moment I Realized I Hate War as Only a Warrior Can, Head Space and Timing Blog, 18 October 2016, http://veteranmentalhealth.com/the-moment-i-realized-i-hate-war-as-only-a-warrior-can/

40. Episode 117 of the Head Space and Timing Web site, Sebastian Junger – Tribe and the New Veteran Homecoming, http://veteranmentalhealth.com/hst117/

41. Winslow Homer was an American artist, known as one of the foremost painters in 19th Century America. His painting titled "The Veteran in a New Field" was painted soon after General Robert E. Lee's surrender on April 9, 1865, and President Abraham Lincoln's assassination five days later. The canvas depicts an emblematic farmer, revealed to be a Union veteran as well by his discarded jacket and canteen at the lower right. https://www.metmuseum.org/toah/works-of-art/67.187.131/

42. Ibid

43. The Wounded Healer and Veteran Mental Health, Head Space and Timing Blog, 16 May 2017 http://veteranmentalhealth.com/the-wounded-healer-and-veteran-mental-health/

44. Comprehensive Veteran Mental Health Addresses All Aspects, Head Space and Timing Blog, 22 February 2018, http://veteranmentalhealth.com/comprehensive/

45. Lack of Access, Cost, and Stigma Continue to Prevent Veterans & Families from Receiving Mental Health Care, Cohen Veterans Network Web site, 7 August 2018 https://www.cohenveteransnetwork.org/wp-content/uploads/2018/08/Barriers-to-Care-August-2018.pdf

46. Through the Other Side of the Valley of Death: Veterans and Posttraumatic Growth, Head Space and Timing Blog, 9 March 2016 http://veteranmentalhealth.com/through-the-other-side-of-the-valley-of-death-veterans-and-posttraumatic-growth/

47. Resistance or Resilience? Head Space and Timing Blog, 6 April 2017, http://veteranmentalhealth.com/resistance-or-resiliance/

48. 2015 Demographics: Profile of the Military Community, Department of Defense, https://download.militaryonesource.mil/12038/MOS/Reports/2015-Demographics-Report.pdf

49. 6 Facts About the U.S. Military and Its Changing Demographics, Pew Research Center Web site, 13 April 2017 https://www.pewresearch.org/fact-tank/2017/04/13/6-facts-about-the-u-s-military-and-its-changing-demographics/

50. These States Have the Highest (and Lowest) Enlistment Rates in America, James Clark, Task & Purpose, 27 June 2017, https://taskandpurpose.com/states-highest-lowest-enlistment-rates

51. Nontraditional Students Gaining Steam in Higher Ed Discussions, Patti Zarling and Shalina Chatlani, Education Dive Web site, 14 November 2017, https://www.educationdive.com/news/nontraditional-students-gaining-steam-in-higher-ed-discussions/510865/

52. Underemployment: Research on the Long-Term Impact on Careers, Burning Glass Technologies Web site, https://www.burning-glass.com/research-project/underemployment/

53. Junger, S. (2016). *Tribe: On homecoming and belonging.* Twelve.

ABOUT THE AUTHOR

Duane France is a retired U.S. Army Noncommissioned Officer, combat veteran, and clinical mental health counselor practicing in Colorado Springs, Colorado. In addition to his clinical work, he also writes and speaks about veteran mental health to a wide audience through his podcast and blog, Head Space and Timing, which can be found at www.veteranmentalhealth.com

Made in the USA
Monee, IL
10 June 2020